Praise for

STEALING GOD'S THUNDER

"Delightful . . . Dray offers a survey of Franklin's scientific career, describing both the ridicule and glory that his experiments inspired." —*The Wall Street Journal*

"Absorbing [and] stylishly written . . . There are other Franklins—the entrepreneur, the diplomat, the statesman, the architect of independence—but in Franklin the scientist, Mr. Dray may have found the happiest one of all." —*The New York Times*

"[Dray's] prose is elegant [and] Franklin is charming." —*The New York Times Book Review*

"A masterful glimpse of . . . Franklin's work . . . a captivating cultural history of Franklin's America . . . [and a] lively and entertaining tale." —*Publishers Weekly* (starred review)

"[An] illuminating study . . . elegantly written." —*Los Angeles Times*

"An entertaining, revealing picture of Franklin as a scientific revolutionary." —*San Jose Mercury News*

"Timely. Dray makes clear that Franklin brought to his political work the same rationalism that informed his science. . . . [Written] with energy and economy. [Dray's] treatment of the electrical investigations . . . is the fullest currently available." —*The Washington Post Book World*

"Benjamin Franklin appears to be in the ascendance as one of the most interesting and accessible Founders. Dray's new entry in the Franklin sweepstakes is elegantly written, mercifully free of the scholarly jargon Franklin would have made fun of but wise and scholarly in the best sense of the term. *Stealing God's Thunder* strikes me as the best study of Franklin as a scientist ever written."
—Joseph J. Ellis, Pulitzer Prize–winning author of
Founding Brothers: The Revolutionary Generation and
His Excellency: George Washington

"Tracing Franklin's beliefs through science, Dray's congenial history has information that will surprise even veteran Franklin fans."
—*Booklist*

"Informative and engaging. Dray highlights Franklin's intense curiosity about nature, his penchant for clarifying the mystery and chaos of natural phenomena, and the particular bent of mind that allowed him to understand previously unrecognized cause-and-effect relationships in the world of the physical sciences. . . . By book's end we see Franklin as a paragon of the American inventive and creative spirit."
—*Orlando Sentinel*

"[Written] with expertise, wit, and balance. After you read [this] book, you'll know a new Ben, and you'll never again take your light switches for granted."
—*Virginia Quarterly Review*

"Dray believes that Franklin's discoveries are a metaphor for the Enlightenment movement, the symbol of revolution and a 'model for human liberation.'"
—*Deseret Morning News*

"Philip Dray captures the genius and ingenuity of Franklin's scientific thinking and then does something even more fascinating: He shows how science shaped his diplomacy, politics, and Enlightenment philosophy."
—Walter Isaacson, author of
Benjamin Franklin: An American Life

"Philip Dray has coaxed the familiar toward new dimensions and has succeeded in making the complex entirely, enthrallingly clear. This is a wise and lucid book, vastly informative, and a pleasure to read." —STACY SCHIFF, Pulitzer Prize–winning author of *Véra: (Mrs. Vladimir Nabokov)* and *A Great Improvisation: Franklin, France, and the Birth of America*

"To the familiar Franklin as writer, printer, politician, and diplomat, Philip Dray adds a marvelous portrait of Franklin as scientist, justly acclaimed in his own day for his innovative study of electricity. A well-told tale that will interest readers of all descriptions." —MARY BETH NORTON, author of *In the Devil's Snare: The Salem Witchcraft Crisis of 1692*

"We forget, living in this era of heavily patented research and closely guarded results, how wonderfully exciting the scientific world used to be. In *Stealing God's Thunder,* the story of Benjamin Franklin's invention of the lighting rod and the resulting consequences, that sense of wonder and excitement and even fear comes beautifully to life. Philip Dray does a remarkable job of illuminating the ever fascinating Franklin and, more than that, the way that he, and his invention, helped create the new scientific world." —DEBORAH BLUM, author of *Love at Goon Park: Harry Harlow and the Science of Affection*

ALSO BY PHILIP DRAY

At the Hands of Persons Unknown:
The Lynching of Black America

STEALING

 GOD'S

THUNDER

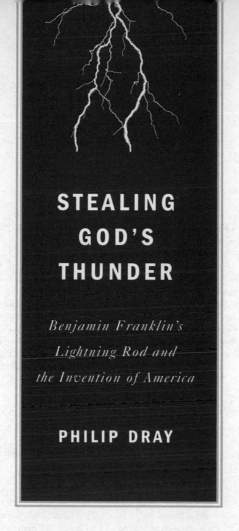

STEALING GOD'S THUNDER

*Benjamin Franklin's
Lightning Rod and
the Invention of America*

PHILIP DRAY

RANDOM HOUSE TRADE PAPERBACKS | NEW YORK

2005 Random House Trade Paperback Edition

Copyright © 2005 by Philip Dray

Published in the United States by Random House Trade Paperbacks,
an imprint of The Random House Publishing Group,
a division of Random House, Inc., New York.

RANDOM HOUSE TRADE PAPERBACKS and colophon are trademarks
of Random House, Inc.

Originally published in hardcover in the United States by
Random House, an imprint of The Random House Publishing Group,
a division of Random House, Inc., in 2005.

ILLUSTRATION CREDITS: The illustrations are from The Franklin
Collection of the Yale University Library, from the Picture
Collection at the New York Public Library (Mid-Manhattan
Branch), or from the author's collection. The woodcut images,
which appear by permission of The Franklin Collection, are
from *Benjamin Franklin: A Biography in Woodcuts* by Charles
Turzak, with text by Florence Turzak (Covici Friede, 1935).
The photograph on page 67 is by Paul Strand and appears
in *Time in New England*, photographs by Paul Strand,
text selected and edited by Nancy Newhall
(Aperture Books, New York: 1980).

LIBRARY OF CONGRESS CATALOGING-IN-PUBLICATION DATA
Dray, Philip.
Stealing God's thunder: Benjamin Franklin's lightning rod and
the invention of America / Philip Dray.
p. cm.
Includes bibliographical references and index.
ISBN 0-8129-6810-7
1. Franklin, Benjamin, 1706–1790. 2. Franklin, Benjamin,
1706–1790—Knowledge—Physics. 3. Electricity—
Experiments—History—18th century.
4. Lightning—Experiments—History—18th century.
5. Statesmen—United States—Biography. 6. Physicists—
United States—Biography. I. Title.
E302.6.F8D69 2005
973.3'092—dc22
[B]   2004051349

Printed in the United States of America

www.atrandom.com

2 4 6 8 9 7 5 3 1

In the Affairs of this World,

Men are saved,

not by Faith,

but by the Want of it.

POOR RICHARD'S ALMANAC, 1754

CONTENTS

INTRODUCTION

ONE DAY IN APRIL 1755, BENJAMIN FRANKLIN WAS RIDING ON A country lane in Maryland with his son, William, and some other gentlemen, when he suddenly saw in the valley below "a small whirlwind beginning in the road, and showing itself by the dust it raised." Franklin spurred his horse as the whirlwind—at its base no larger than a barrel, but as much as thirty feet wide at the top—approached the group. "The rest of the company stood looking after it, but my curiosity being stronger, I followed it, riding close by its side. . . . As it is a common opinion that a shot, fired through a water-spout, will break it, I tried to break this little whirlwind, by striking my whip frequently through it, but without any effect."

Franklin chased the whirlwind into a wooded area, where it began picking up leaves. "I accompanied it about three quarters of a mile, till some limbs of dead trees, broken off by the whirl, flying about and falling near me, made me more apprehensive of danger;

and then I stopped, looking at the top of it as it went on, which was visible, by means of the leaves contained in it, for a very great height above the trees." Franklin rode back to join his companions, who had halted in the road to enjoy watching his pursuit. "Upon my asking Colonel Tasker if such whirlwinds were common in Maryland, he answered pleasantly: 'No, not at all common; but we got this on purpose to treat Mr. Franklin.' "

"The rest of the company stood looking after it, but my curiosity being stronger, I followed it, riding close by its side." This comment can function, suggests biographer Carl Van Doren, as a declaration of Franklin's scientific philosophy: where most people gave, at best, only passing thought to unexplained natural phenomena, Franklin was endlessly intrigued. Whirlwinds and waterspouts, storms, the principles of heat and ventilation, oil's calming effect on water, the mysterious Gulf Stream, electricity that "enters and leaves its temporary homes, always returning to itself unaltered"—these were among the subtle actions of nature for which he maintained a life-long fascination, from his boyhood experiments with swimming "palettes" in a Boston mill pond to his delight, decades later as a statesman in Paris, in the novelty of balloon flight.

Franklin's success, as a diplomat and one of our nation's founders, is often credited to his supreme cool—his ability to remain above the fray, to act in the broad interest of conciliation, to say only what was helpful or required and no more. Science was always a different story. With its indifference to sentiment and its insistence on observable fact, it was for Franklin an ideal foil. "Nature alone met him on equal terms," historian Carl Becker has said, "with a disinterestedness matching his own. . . ."

Of course, in an age when so much remained unknown about the workings of the physical world, prying into nature's secrets could be risky. In December 1750, Franklin had invited some friends to his Philadelphia home to help him slaughter a turkey by electrocution for Christmas dinner, using a charge from two Leyden jars. An experienced printer, Franklin had powerful hands and was extremely dexterous. But, in this instance, allowing himself to be distracted by his guests, and not minding for a moment what surfaces he was touching, he "inadvertently took [the stored electricity] through my own arms and body, by receiving the fire from the united top wires with one hand while the other held a chain connected with the outsides of both jars. The company present . . . say that the flash was very great and the crack as loud as a pistol. I then felt what I know not well how to describe: a universal blow throughout my whole body from head to foot, which seemed within as well as without." One of Franklin's hands, "feeling like dead flesh," instantly turned white, as if all the blood had been driven out. He may have had this experience in mind when he later speculated, a bit naïvely, that the effect of being struck by a lightning bolt might be approximated by wiring up several Leyden jars.

Even so near a brush with death did not inhibit his interest in electrical experiments or diminish his essential optimism. "We learn by chess," observed Franklin, a dedicated player of the game, "the habit of not being discouraged by present appearances in the state of our affairs, the habit of hoping for a favorable change, and that of persevering in the search for resources."

Franklin's inquisitiveness about nature was matched by his disdain for the continual upheaval he found in human society. Much of his life, as a printer, publisher, and inventor, as a civic leader, and as a statesman, was spent trying to replace chaos with order—creating a program to keep the streets clean, a volunteer fire department, a young men's improvement society, a lending library, a plan to ally the American colonies. It was probably a union of these two impulses—his interest in investigating that which had not yet been adequately explained, and the urge to fix what could bear improvement—that drew his attention initially to the phenomena of thunder and lightning and led to his greatest invention, the lightning rod.

Thunderstorms "inflam'd my Curiosity," he once said. Representing nature on a vast scale and at its most frightening and enigmatic, storms became a particular subject of interest to Franklin during his early years in Philadelphia in the 1730s, when he was publisher of the *Pennsylvania Gazette* and was beginning his public life. Little or nothing was known about lightning, whose true nature was then shrouded by centuries of superstition. When, in summer, lightning flickered over the New Jersey shore of the Delaware River, Franklin paused to watch its zigzag path across the sky. He took note of the manner in which it destroyed trees and chimneys, and calculated its downward course to the ground from the tops of church steeples. He filled the *Gazette* with accounts of the "mischief of thunder-gusts," and later—to the dismay of his wife, Deborah—rigged an apparatus that would bring lightning into their house and announce its arrival by tinkling some bells on the stairway landing.

"[On] Tuesday last the House of Mr. James Rogers was struck by Lightning," Franklin reported in his paper on July 10, 1732.

It split down Part of the Chimney, went through the Room where he was sitting with his Children, but without hurting any of them; and entring into the Cellar, fir'd a full Hogshead of Rum which stood under an Arch, and bursting out the Head, the whole Cellar was instantly fill'd with Flames which pour'd out at the Windows. There was several hundred weight of Butter in Tubs, which melted and took fire also. . . . After the Fire was out, it was so hot as not to be tolerable to the Feet and Legs of those who would have gone in.

But "mischief" could also connote more dire effects:

From Newcastle we hear that on Tuesday the 8th instant, the lightning fell upon a house within a few miles of that place, in which it killed 3 dogs, struck several persons deaf, and split a woman's nose in a surprizing manner.

In the report of another incident, Franklin the journalist takes up a subject that would one day captivate Franklin the scientist— the way lightning, in a tiny fraction of a second, chooses its route through the available conductors in an edifice it has struck:

On Sunday the 28th past, about three Miles from this City, a Clap of Thunder fell upon the House of the Widow

Mifflin, struck down part of the Top of the Chimney, and split it down several Feet; tore and shattered the Roof, split a Rafter, and broke it off in two Places: and struck off the Plaistering with part of the Brick Wall in the inside of a lower Room, broke the Splinters of the Window Frame and melted the Lead; another Story lower several Splinters of the wood frame were broke off, some of the Glass broke, and Lead melted; and a Lad who stood in a Porch near the Window, was struck down, and burnt badly in a Streak about the Breadth of one's Hand, from the Side of his Face down to the Calf of his Leg; but in no way hurt any part of his Clothes.

Franklin included lightning stories in his paper because they made for interesting, sometimes humorous, copy. At times, he almost humanized the thunderbolts, as if they were reckless or prankish children. "Mischief"—the word he used to describe lightning's destructive nature—was the same term he applied to human situations he thought both regrettable and avoidable: the theft of a horse left untethered, the disruption in a household caused by a runaway servant, the animosity between residents and billeted British troops in his native Boston. No wonder, then, that his lightning rod was ultimately said to have "tamed" lightning.

Because of his work with electricity, Franklin was popularly viewed, especially in England, where he served as colonial agent, and later in France, where he was foreign minister for the burgeoning United States, as a kind of genial sorcerer. But many of his scientific contemporaries saw him in more exalted terms, and acclaimed him as another Galileo, Newton, or Copernicus. Franklin was not the deep theorist these men had been, but his lightning rod—some called them "Franklin rods"—represented the fulfillment of two important strands of eighteenth-century scientific thought: the belief that science should be based on empiricism, that which could be observed, and the confidence that an empirical approach to science would lead to the development of innovations beneficial to mankind. What made Franklin's achievement especially remarkable in European eyes was that he was an American, at a time when natural philosophers across the Atlantic looked to the New World for botanical curiosities and snake skins and not much else, when the very term "American" still connoted a coarse frontiersman.

"The usual portrayal of Franklin presents him as a political figure who, in his spare time, dabbled in science," the historian I. Bernard Cohen has said. "His own century, on the other hand, considered him a scientist who had entered the arena of international politics." Diplomat, publisher, satirist, revolutionary, Franklin did many things well. Future generations came to know him best—thanks in large part to his own posthumously published *Autobiography*—as the son of a Boston candle-maker who, by dint of ambition, hard work, and self-discipline, rose to world fame and influence in spite of what he called his "low beginning." As Cohen suggests, however, what has always been inadequately acknowledged is that Franklin's initial world fame rested not on his wit or his accomplishments in public affairs: It was based on his discoveries in electricity and his invention of one of the most useful and symbolically important conveniences in human history.

Through Franklin's lightning rod and his leading role in fostering American independence, mankind found in him both a philosopher who banished superstition and a revolutionary who humbled kings. "He snatched lightning from the sky and the scepter from tyrants," the French economist Anne-Robert-Jacques Turgot famously observed of America's "electric ambassador" to the court of France. In this way, Franklin and his lightning rod became a popular metaphor for the late Enlightenment and the emerging faith that society, by emulating science's rationalism, would itself be transformed. The eighteenth century's iconography of revolution is replete with examples of knowledge and liberation transmitted by electric spark. Just as Michelangelo's God passes the gift of life to Adam with his fingertips, Franklin, in a celebrated painting by Benjamin West, returns to confront the Almighty. Surrounded by cherubic helpers, his white hair whipping in the storm, Franklin uses a kite and a key to gather electricity from a thunderbolt, bringing to earth the light of Reason and sparking the modern age.

Today, it is difficult to appreciate how dramatically innovative

Franklin's ideas were in 1750, when electricity itself was barely understood. His proof that the august powers of thunder and lightning were, in reality, only a variation of the "spark and snap" experiments he conducted on a tabletop in his Philadelphia home constituted a major scientific breakthrough. But it was also much more. For its cultural significance, this discovery's impact on the eighteenth century has been compared to that of the atom bomb in the twentieth, and described as a moment in human history as epochal as the birth of Christ.

The lightning rod, however, did not become so encouraging a beacon to mankind without first meeting fierce resistance. The idea that either a resentful God or "diabolical agency" directed lightning bolts was a deeply ingrained belief; violent storms, it was thought, could best be dispelled by the ringing of specially "baptized" church bells and the mumbling of prayers. Lightning rods, which reinterpreted, or perhaps even voided, the relationship between God and his human supplicants, were thus heretical. In 1755, Boston's clergy went so far as to accuse Franklin of being responsible for a destructive New England earthquake, for surely the rods were acting to deflect heaven's "resentment" into the ground. The device thus raised questions the world was at first unsure of how to answer: Did humanity have the *right* to defend itself from threatening weather, disease, or other calamities long deemed providential? If so, how was such a new insight about man's relationship with nature to be assimilated into the sustaining power of faith? And if lightning really was a natural force, what *was* it exactly, and could it be safely scrutinized and measured?

In England, Franklin and his invention became enmeshed in a bitter scientific controversy involving King George III. Around the same time, in rural France, a sensational court case over a man's right to safeguard his house from lightning brought to the fore a brash young attorney, Maximilien de Robespierre. And late in life the famous Franklin, having given the world a rod that "governed" thunder, was handed the dangerous public assignment of passing judgment on a rod that appeared to control human health and well-being, one wielded by a man whose popularity in Europe rivaled his own—the charismatic Franz Anton Mesmer.

Two and a half centuries after Franklin first installed a lightning rod on a Philadelphia rooftop, the Enlightenment's special reverence for it as an icon of change is largely forgotten. Yet the questions it raised—of reason and faith, liberty and tyranny, science and superstition—persist, while the spark at the lightning rod's tip, glowing from Franklin's century to ours, reminds us that we must sometimes be willing to confront those things we fear most.

Curiosity, courage, and their rewards, then, are the subjects of this book: how a man with no formal scientific training, asking questions and feeling his way as he went, probed the enigma of electricity and wound up explaining it; how his approach to science enabled him to accomplish so much in other spheres; and how the lightning rod became a model for human liberation, and helped make Franklin's life, and the founding of the United States, two of the Enlightenment's exemplary attainments.

STEALING

 GOD'S

THUNDER

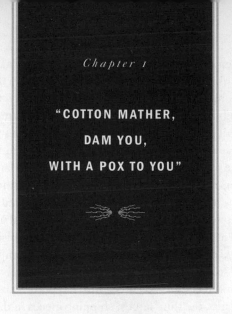

Chapter 1

"COTTON MATHER, DAM YOU, WITH A POX TO YOU"

BENJAMIN FRANKLIN WAS BORN IN BOSTON ON JANUARY 17, 1706, THE fifteenth of the seventeen children of Josiah Franklin, and the eighth child of his father's second wife, Abiah Folger. The Franklins lived on Milk Street, across from the South Church, where Josiah was a leading member of the congregation. Ben was carried across the street and baptized there on the day of his birth. The Franklins ran a soap- and candle-making business, and Josiah was also active in the community; he had served as a constable of the town watch and also in the public markets, neighbors sometimes came to him for advice, and the son would recall that his father's "great Excellence lay in a sound Understanding and solid Judgment in prudential Matters, both in private and publick Affairs."

While still a toddler, Ben struck his parents as having the bearing of a scholar. "I do not remember when I could not read,"

Franklin later said. An uncle (also named Benjamin) who resided with the Franklins, and who took a special interest in his namesake, perceived something remarkable about his clever nephew, and wrote of the boy, "If the Buds are so precious what may we expect when the fruit is ripe?" Josiah prided himself that his youngest son might possess the makings of a clergyman, although Ben's unsuitability for the role manifested itself early on in ways large and small, such as when he suggested to his father that if all the meat being salted for the family's winter provisions was blessed at once, the family might avoid having to say grace at each meal and "it would be a vast saving of time." In any event, the prerequisite education for the clerical calling proved too costly, and after completing barely two years of school, Ben was put to work in the family shop.

Boston in the early 1700s was a thriving port of about ten thousand inhabitants, the third largest shipping mecca in the British empire, with fifteen shipyards and hundreds of wharves that teemed night and day with the loading and off-loading of goods and passengers. Ben was smitten with the magnificent sight of ships—the packets, cargo vessels, and men-of-war that stood shoulder-to-shoulder at the docks, their vast sails filling the sky. The town's seafaring character, with its inlets, rivers, bays, ponds, and coves, engendered in him a lifelong affection for boats and the sea. "Living near the water, I was in and about it, learnt early to swim well, and to manage boats," he recalled. As the adolescent leader of a ragtag army of boys who played around the large mill pond that began just beyond his father's shop, Ben became dissatisfied with the speed he could obtain through his regular swimming strokes and experimented with ways to improve his efficiency by attaching "palettes" (flippers) to his hands and feet. Already a deft observer of the movements of air, water, and wind, he also conceived of a most uncommon experiment, flying a kite while submerged in the water. "Having then engaged another boy to carry my clothes around the pond, I began to cross the pond with my kite, which carried me quite over without the least fatigue and with the greatest pleasure imaginable." As the kite drew him swiftly from one side of the pond to the other, a band of excited youngsters ran along the shore, shouting and encouraging his progress.

One of Josiah Franklin's other sons, also named Josiah, had been lost at sea, and the father, concerned about Ben's evident fondness for ships, sought to head off any seafaring inclinations the young boy might have. But Ben was clearly apathetic about work in the family trade, and the candle shop was not without dangers of its own: the boiling vat Ben was made to stir was the very one in which another brother, an infant named Ebenezer, had earlier drowned. "Under Apprehensions that if he did not find [a job] for me more agreeable, I should break away and get to Sea, as his Son Josiah had done to his great Vexation," Franklin later wrote, Josiah determined to establish Ben in a trade. After a brief, unhappy turn at a cutlery shop, Ben signed papers of apprenticeship to a printing business owned by his older brother James.

James Franklin, nine years older than Ben, was a worldly so-phisticate by Boston standards. In 1717 he had visited London, where he had been inspired by English satirists such as Joseph Addison, the Irish-born Jonathan Swift, and Daniel Defoe, and by the chatter of anti-authoritarian opinion in London's Grub Street coffeehouses. James probably knew his kid brother and new ap-prentice only marginally, although it immediately became clear that Ben possessed the attributes necessary for work in a print shop—manual dexterity, mental quickness, and physical strength. As Ben would remember, "In a little time I made great proficiency in the business, and became a useful hand to my brother." He also dipped hungrily into James's library, devouring books on arith-metic and navigation as well as the many pamphlets and sermons James printed that argued prevailing notions of theology, Deism, and natural philosophy. He read the third Earl of Shaftesbury's writings urging moderation in human affairs and an end to religious zealotry, John Locke's *An Essay Concerning Human Understanding*, Daniel Defoe's *Essay on Projects*, and Cotton Mather's *Bonifacius: An Essay Upon the Good*—books that "perhaps gave me a Turn of Thinking that had an Influence on some of the principal future Events of my Life."

In these books, Ben discovered a world of ideas to absorb and act upon. By the time he was a teenager, he had declared himself a vegetarian, perfected a "scientific" method of swimming, adopted

Socratic inquiry as a means of arriving at truth, and embraced Deism, the belief that the world had been made by a gifted supreme being who was no longer a living feature of the cosmos. This *deus absconditus*, apparently pleased with his creation, had left it to its own devices; the Deists believed that God's endowment to mankind was reason and the practice of benevolence toward others, and that Truth resided not in biblical aphorism but in nature.

The age in which Ben Franklin grew up, the early Enlightenment, was a time of expanding faith in individual experience and self-determination, an era of enlarged human curiosity, when advances in natural philosophy, science, and technology were curbing man's reliance on magical or religious explanations for life's hardships. Like all such epochs of cultural transition, it arose from sources both subtle and diverse, but certainly the English philosophers of the seventeenth century and their Scientific Revolution had helped set the stage. Its founding concepts lay in the ideas of Francis Bacon, who promoted the value of experimentation and promised that work in pure science would lead to useful applications, and those of John Locke, who urged that true knowledge came not from a perfected obedience to inherited beliefs but from mankind's environment, as well as from the instruments that made new forms of observation possible, such as the microscope.

Hovering majestically above all was Isaac Newton, the guiding scientific thinker of his day, who showed in his *Philosophiae Naturalis Principia Mathematica*, published in 1687 and known as the *Principia*, that the functions of the universe were governed by mathematics. Copernicus had made the sun, not earth, the center of the universe; Johannes Kepler described the planets' orbits as ellipses; but where others glimpsed a corner of the tapestry, Newton saw that the whole—earth and heavens alike—was a single system, bound by physical laws. "The new [Newtonian] mechanics was the first example of a modern science in its full development," historian Dirk J. Struik notes, "equipped with a convincing set of axioms, a logical method, a developed technique and the power of forecasting events. Even those who could not master the details could admire and follow its general approach." As Voltaire, who helped introduce Newton's ideas in France, observed, "Very few people read Newton, but everybody talks about him."

Such reorganization of human thought did not come altogether easily to the Boston of young Ben Franklin. New Englanders still lived to a great extent in "an enchanted universe"—a place of dark, impenetrable forests, vengeful thunderstorms, portentous comets, witches, and ghosts. A belief in providential events and omens was common, as was the idea that an ongoing battle between Satan and God ruled many features of daily life. For people who suffered from myriad poorly understood health issues, whose children died in great numbers, and who had no other means of comprehending the phenomenal world, providence and its signs offered at least the surety that man's earthly trials had some purpose. "In place of unacceptable moral chaos," writes the scholar Keith Thomas, "was erected the edifice of God's omnipotent sovereignty," a presence that seemed particularly reasonable given the Puritan assumption that God had a special interest in the fate of New England.

"Without doubt the Lord Jesus hath a peculiar respect unto this place, and for this people," preached the influential Boston cleric Increase Mather. "Christ by a wonderful Providence hath dispossessed Satan, who reigned . . . in these Ends of the Earth for Ages . . . and here the Lord hath caused . . . New Jerusalem to come down from Heaven."

This "peculiar respect," however, contingent on the Puritans' steadiness of faith, and known as the New England or Puritan covenant, seemed depleted of late. A bloody conflict with the Indians in 1675–78, known as King Philip's War, had demoralized Massachusetts, Maine, Connecticut, and Rhode Island, and forced the settlers to recognize that they were capable of violent savagery equal to any the "savages" could offer. Fires among Boston's wooden houses and narrow lanes had twice turned the city into a tinderbox, and in 1678 a deadly smallpox epidemic had arrived. In August 1682 a bright, portentous sign (soon to be known as "Halley's comet") crossed the New England sky, the mysterious celestial visitor raising what Increase Mather called "Heaven's Alarm to the World." In 1684 the long-standing charter of the

Massachusetts Bay Colony was abrogated by the English crown, diminishing the hegemony of the Puritans and stirring political uncertainty. With disillusionment spreading, clergymen worried about increased Sabbath-breaking and even whispered about a rise in bestiality, a trespass so Godless it was punishable by death. Finally, in 1692–93, the infamous Salem witch trials occurred.

In the witchcraft cases, twenty individuals were put to death, many on what was known as "spectral evidence": confessions by those who said they were afflicted by specters, or accusations by people who claimed to have seen their neighbors assume the forms of devils and witches. Cotton Mather, Increase's son and a ranking clergyman in his own right, was asked by the judges to write a defense of the need for the trials. Cotton made a nominal argument against the future reliance on spectral evidence, but he was unable to keep himself from defending the Puritan authorities as he lambasted others for Sabbath-breaking, adultery, and drunkenness. Public unease over the executions, and new doubts about the legality of spectral evidence as proof, helped lessen the likelihood of further witch trials, but Mather's arrogant, scolding rant was long remembered.

Ben's introduction to the new intellectual currents crossing the Atlantic, gained among his brother's books, was timely, for in the spring of 1721 a crisis arrived that would test some of these principles and plunge Boston, the Mathers, and both Franklin brothers into a maelstrom.

Men and women of the early eighteenth century called smallpox "the speckled monster." Of all illnesses it was perhaps the most horrific, turning its victims into hideous sufferers, their limbs scarred, their faces suppurating with diseased pustules. Several earlier outbreaks in Boston had led to strict rules for the quarantining of incoming ships and their crews, but there had been no local epidemic since 1702, and by 1721 many of the former precautions had been allowed to lapse. Thus an entire generation had grown up with no exposure to the disease when, on April 22, an epidemic rode into Boston Harbor aboard the H.M.S. *Seahorse,* en route from the West Indies. Within a few weeks the first cases of "the pox" were reported. Panic spread quickly across the Charles River to Cambridge, where Harvard canceled its June commencement exercises. As the disease claimed more victims, Governor Samuel Shute called for a day of fasting and humiliation on June 27 to combat whatever misdeeds "have stirred up the Anger of Heaven against us."

No one watched these events more closely than Cotton Mather, the minister at Boston's North Church. In 1713, Mather had written the first layman's guide to the treatment of measles, only to see his second wife, Elizabeth, three of his children, and his maid succumb to the disease—all in the same month. Mather had already lost his three eldest children to illness in 1687, 1696, and 1699, and his first wife, Abigail, had died in the smallpox epidemic of 1702. In total, he would lose two wives and thirteen of his fifteen children to disease (or, in one case, a birth defect) and be survived by only two, Samuel and the roguish Increase, Jr., known as "Creasy," who had recently devastated his father by dropping out of Harvard and "getting an Harlot big with a Bastard."

For some time Mather had nurtured an interest in a little-known method of disease prevention known as inoculation, which he had first learned of from his slave, Onesimus, who had come to

Mather in 1707 as a gift from his congregation. Other slaves confirmed Onesimus's claim that, in Africa, inoculation—inserting a small amount of diseased material into a vein to promote a mild case of smallpox so a greater one would be avoided—was practiced and had proven effective. In 1713 Mather had become one of the first Americans elected to the Royal Society of London, the most prestigious scientific organization in the world. (Its founding in 1662 was inspired by Bacon's tenet that science be a collaborative undertaking. The Society's motto, *"Nullius in verba"* ["On the word of no one!"], honored the idea that science must rest on observation.) Cotton sent eighty-two letters describing American natural curiosities and his experiments in corn hybridization to the Society between 1712 and 1724.

Mather saw not dissonance in the twin callings of science and faith, but a powerful unification. "[Natural] Philosophy is no Enemy, but a mighty and wondrous Incentive to Religion," he wrote. "The Whole World is indeed a Temple of God, built and fill'd by that Almighty Architect."

In 1699 the Royal Society had learned of early inoculation procedures in China, and beginning in spring 1714, accounts of inoculation began to appear in its publication, the *Philosophical Transactions*. A report by Dr. Emanuel Timonius, a graduate of Oxford, was paired in an edition of 1716 with an account by Dr. Jacobus Pylarinus, who had learned that inoculation was being practiced in Turkey. Mather also may have heard reports of Lady Mary Wortley Montagu, wife of the British consul in Constantinople, who reported in 1717 that Turkish peasant women held "smallpox parties" for "engrafting" others with a minute amount of smallpox matter. In England, inoculation, known as "buying the small pocks," was tested on prison inmates and began to gather adherents, although not quickly enough for Mather. "How does it come to pass that no more is done to bring this operation into experiment and into Fashion in England?" he demanded of a London correspondent in 1716. "For my own part, if I should live to see the Small-Pox again enter into our city, I would immediately procure a consult of our Physicians, to Introduce a Practice [inoculation], which may be of so very happy a Tendency."

Mather was one of the few people in North America who had examined pustules of smallpox through a microscope, and even as from his pulpit he continued to deliver fiery jeremiads about sin and the providential calamities that awaited nonbelievers, he quietly came to embrace a most rational notion, later to be known as "germ theory," the idea that pathogenic animalcules carry disease.

"Every part of matter is peopled," he wrote. "Every green leaf swarms with inhabitants. The surfaces of animals are covered with other animals. Yea, the most solid bodies, even marble itself, have innumerable cells, which are crowded with imperceptible inmates." These tiny animals, breathed in by humans, he believed, could attack the body's organs. By receiving inoculation in the arm, however, a person gave his body's natural defenses an opportunity to confront the invader at a safe, manageable remove, before the organs could be reached.

When the 1721 outbreak occurred, Mather was ready to act on his vow to inoculate smallpox sufferers. He was, however, a divisive character to be advocating so controversial and unheard-of a practice. Overbearing and full of his own sense of destiny, he claimed to have once conversed with an angel, and was so consumed with holiness it was said he implored God's grace even while urinating. The son of the town's most well-known clergyman, an adolescent who had wrestled with and ultimately conquered a severe speech impediment, he was the kind of person who senses they are not well-liked yet cannot seem to avoid alienating others. Bostonians respected and probably feared him a bit, even as they found his personality intolerable.

In early June, when Mather wrote to suggest a program of inoculation to the members of Boston's "medical community," a loose fraternity of local learned men who, like Mather, dabbled at "physick," opposition was voiced immediately by the town's only academically trained physician, William Douglass. An opinionated Scotsman, Douglass was known for his hatred of the French, the Indians (he called them "manbrutes"), and most forms of civil or religious authority; like many people, he found the notion of curing someone by intentionally giving them a deadly disease counterintuitive. He also had little patience with nonprofessional medical practitioners, or "empyrikal quacks," as he called them, and refused to be impressed by Mather's scientific pretensions, openly doubting his membership in the Royal Society and lampooning a theory Mather had once published suggesting that certain migratory birds, departing New England in the fall, "repair to some undiscovered satellite" for the winter.

Douglass's criticisms were a bit unfair. Due to the lack of trained physicians in the New World, it was common for clergy, midwives, ship's surgeons, or anyone with a calming touch to attend to the sick and dying. And Mather's eclectic interests were not unusual in a time when any natural philosopher worth the name was drawn to a wide range of subjects, from botany to medicine to astronomy.

Mather did have one ally in Zabdiel Boylston, a minister who, like Mather, had survived the 1702 epidemic. On June 26, 1721, Boylston inoculated his six-year-old son, Thomas, a thirty-six-year-old male slave named Jack, and the latter's two-and-a-half-year-old son, Jacky, conducting the first known experiment in inoculation in the New World. Boylston soon reported that although his son suffered a fever as a result of the inoculation, the patients appeared to be doing well (all three survived the epidemic). "No one need fear, in this way, of having many Pustules, of being Scarred in their Face, or of ever having the Small-Pocks again," Boylston wrote in the *Boston Gazette*.

However, a week later, on July 21, the town selectmen, unsure of inoculation's medical value and concerned that white men should so readily accept a method based on "negroish evidence," ordered Boylston to stop the inoculating. "It is caviled . . . that this New Way comes to us from the Heathen, and we Christians must not Learn the Way of the Heathen," Mather noted privately. "Gentlemen Smoakers, I pray, whom did you learn to Smoke of?" But in the *Boston News-Letter* of July 24, William Douglass charged that the practice was reckless, calling Boylston incompetent and his *Gazette* article "a dangerous quack advertisement." Douglass, who had originally lent Mather the *Philosophical Transactions* describing inoculation, now hotly demanded that Mather return his copies of the journal. On July 31, the Mathers—Cotton and Increase—joined by ministers Benjamin Colman, John Webb, William Cooper, and Thomas Prince, retaliated, publishing a letter in the *Gazette* that praised Boylston's efforts. "Men of Piety and Learning, after much serious tho't," they wrote, "have come into the Opinion of the Safety of the faulted method of Inoculating the Smallpox; and being persuaded it may be a means of preserving a multitude of lives, they accept it with all thankfulness and joy as the gracious Discovery of a kind Providence to Mankind." The public, increasingly disconsolate over the rising number of deaths, became only more confused by the disagreement among the learned men of the town. They were accustomed to being told that God sent disease as righteous punishment; now they were to understand he had sent the remedy.

Ben and James Franklin were to have an exacerbating effect on the dilemma. In August 1721, at the height of the epidemic, James launched a new weekly newspaper, *The New-England Courant*, the first newspaper in colonial America to publicly ridicule and challenge the governing authority. Among the paper's contributors were William Douglass and John Checkley, an Oxford-educated apothecary. The *Courant* immediately became the voice of defiance in the inoculation controversy, sharply questioning the clergy's leadership and Boylston's actions. James mocked inoculation as only the latest in a pattern of clerical high-handedness in New England, from a mid-seventeenth-century assault on the Quakers to Increase Mather's scribblings on the providence of comets, to the fairly recent witch crisis.

Since 1690 and the publication of *Publick Occurrences*, Boston's first newspaper, which had been suppressed by the authorities after one issue, the local press had chosen to err on the side of the stodgy, careful never to give offense. The *Gazette* and the *News-Letter* limited themselves to publishing official proclamations, speeches from Parliament, and weeks-old news from abroad. By contrast, the Franklins' *Courant* was "a vigorous little sheet," featuring local news, editorials, and sometimes biting satire, such as its reprimand to churchgoers who did not sing in harmony or its ad for a fictitious pamphlet entitled "Hoop Petticoats arraigned and condemned by the Light of Nature and Law of God." The Franklins and their colleagues knew they had made a splash when the *News-Letter* went out of its way to denounce the *Courant* as "a Notorious, Scandalous Paper . . . full freighted with Nonsense, Unmannerliness, Railery, Prophaneness, Immorality, Arrogancy, Calumnies, Lyes, Contradictions and what not, all tending to Quarrels and Divisions, and to Debauch and Corrupt the Minds and Manners of New England."

Cotton Mather was particularly disappointed with the Franklin brothers (their father had at times sat in his congregation), while Increase allowed of James that "I cannot but pity poor Franklin, who though but a young man, it may be speedily he must appear before the judgment seat of God." James Franklin in turn denounced the Mathers as "the directors of that wicked practice [inoculation]," along with Boylston ("the Inoculator"), and warned proverbially, "If the blind lead the blind, both fall into the Ditch." Cotton Mather and his "hot-headed Trumpeters," James charged, wanted the *Courant* put out of business so that "he may reign Detractor General over the whole Province, and do all the Mischief his ill Nature prompts him to." Cotton replied that his attackers were only using the inoculation debate as an expedient means of assaulting the clergy. There was some truth to this, although the *Courant*'s actual protest seemed to be that those offering a risky new medical theory should not consider themselves above reproach. In the August 14 issue, Douglass pushed this point, sarcastically proposing that Boylston "inoculate," i.e., murder, the Indians. Increase stepped in, castigating those who opposed his son as "known children of the wicked one." Douglass fired back that Cotton should stick to facts because "railing is not reasoning in this country." Cotton's nephew, the Reverend Thomas Walter, called Checkley an "awkward Flogger." Checkley replied that Walter was a drunk.

And on it went, the nasty brawl of words sending Mather to seek refuge in the pages of his diary. "Warnings are to be given unto the wicked Printer, and his Accomplices," he wrote, "who every week publish a vile Paper to lessen and blacken the Ministers of the Town, and render their Ministry ineffectual. A Wickedness never parallel'd any where upon the Face of the Earth!" Encountering James Franklin in the street, he demanded that the *Courant* stop its editorial attacks, cautioning him that mocking the clergy would surely bring down God's wrath. James, in the next issue of the *Courant*, wrote that what was really bothering Mather was that Bostonians had become more "disposed to give a fairer hearing to what they find in a News-Paper than in a Sermon." In mid-November, a crude grenade was thrown through

Mather's window. The grenade, filled with turpentine and gunpowder, hissed and sputtered but failed to explode, so the master of the house was able to read the attached note: "Cotton Mather, You dog. Dam you: I'll inoculate you with this, with a Pox to you." Mather cited the narrow escape as further evidence of the righteousness of his cause.

Ultimately, of course, Mather *was* proven right. In the Boston smallpox epidemic of 1721, about half the town's residents—5,759—contracted the disease, of whom 842 died. Of the 242 Zabdiel Boylston inoculated, only six died. (Thus, the death rate for those inoculated was 2.5 percent, and the rate for those not inoculated, 14.6 percent.) Boylston later went to England, where he was lionized for his role as a medical pioneer.

Mather was relieved that his program of inoculation was vindicated, but, whether it occurred to him or not, by proving that God could send cures to earth as well as sickness, he had greatly diminished the traditional system of portents and punishments he had spent his life defending. The real lesson he had taught his fellow New Englanders was that it was in no way presumptuous to intercede to alleviate mankind's fear and suffering, and that perhaps God even smiled on such actions. "By introducing into the covenant theory an unprecedented (and worse than that, a not yet explicitly avowed) criterion of utility," historian Perry Miller notes, "[the] inoculates overturned—without quite knowing what they were doing—the corporate doctrine . . . [They] could never again authoritatively contend that what the people suffered was caused by their sins and that repentance alone . . . could relieve them. The clergy themselves had introduced another method, and so brought a fatal conclusion into the very center of their mystique."

Ben Franklin never wrote about the *Courant*'s part in the inoculation battle as he did other aspects of his youth. Not only had the *Courant* been wrong, but the recollection was undoubtedly painful, for in 1736, when he was living in Philadelphia, he lost his son Francis Folger Franklin, known as Franky, age four, to smallpox. Although Franklin by then had come to accept inoculation, he had not bothered to inoculate Franky, possibly because the child was just recovering from another illness. The death crushed both Ben and his wife, Deborah, and for the father it forever remained a sensitive subject. In 1759, while in England as colonial agent for Pennsylvania, he co-wrote with Dr. William Heberden, one of London's most renowned physicians, a booklet explaining inoculation's success rate and urging its use. "Surely parents will no longer refuse to accept and thankfully use a discovery GOD in his mercy has been pleased to bless mankind with," Franklin wrote, and likely thinking of his own misfortune, added, "For the loss of one in ten thereby is not merely the loss of so many persons, but the accumulated loss of all the children and children's children the deceased might have had, multiplied by successive generations."

As the inoculation debate helped weaken the already rickety Puritan leadership, so it also served indirectly as a liberating force in the life of the teenage Ben Franklin. For James Franklin's young apprentice, setting type for the opinionated anti-inoculation articles and hawking the *Courant* on the street, the crisis must have been invigorating, and Ben would have relished the paper's gadfly role in the controversy and enjoyed the tweaking of the imperious Mathers. It was also in the midst of the dispute that he made his own first appearance as a prose stylist. Franklin later remembered in a letter to Samuel Mather:

> When I was a boy, I met with a book, entitled, "Essays to do Good," which I think was written by your father. It had been so little regarded by a former possessor, that several leaves of it were torn out; but the remainder gave me such a turn of thinking, as to have an influence on my conduct through life; for I have always set a greater value on the character of a *doer of good*, than on any other kind of reputation; and if I have been, as you seem to think, a useful citizen, the public owes the advantage of it to that book.

Ben had also been much influenced by a bound volume of the collected issues of *The Spectator*, with essays by Joseph Addison and Richard Steele, that James had in his shop. Published in London in 1711–12, *The Spectator* introduced an innovative style of satire that would be much imitated, criticizing society not with high-handed sermons, but with dry, witty observations. The teenage Franklin devoted hours to copying and memorizing the articles, then rewrote the pieces in his own voice, mimicking their style.

Much of the writing that had appeared in the *Courant* was printed beneath obvious pseudonyms, such as Ichabod Henroost, Abigail Afterwit, Betty Frugal, or Fanny Mournful. Ben began slipping pseudonymous pieces beneath the *Courant*'s door at night, signed "Silence Dogood," whom he introduced as a proper widow who lived in the countryside. From the first installment, appearing April 2, 1722, which includes a memorable description of Mrs. Dogood's "birth" onboard ship en route to New England, "Silence" demonstrates a precocious wit. Her father, raising his arms in exultation at his daughter's safe delivery, is instantly washed overboard, never to be seen again. "Thus," Silence deadpans, "was the first Day which I saw the last that was seen by my Father; and thus was my disconsolate Mother at once made both a Parent and a Widow."

Over the course of thirteen additional Dogood letters, which because of their popularity were soon appearing on the *Courant*'s front page, "Silence" describes her foster father's awkward efforts at courtship, suggests a formula for writing an all-purpose funeral elegy, mocks marriage, clerics, and Harvard students, and holds forth on the importance of free speech. In one entry, "Silence" provides what seems a fair summary of young Ben Franklin himself:

Know then, That I am an Enemy to Vice, and a Friend to Vertue. I am one of an extensive Charity, and a great Forgiver of private Injuries: A hearty Lover of the Clergy and all good Men, and a mortal Enemy to arbitrary Government and unlimited Power. I am naturally very jealous for the Rights and Liberties of my Country; and the least appearance of an Incroachment on those invaluable Priviledges, is apt to make my Blood boil exceedingly . . . To be brief; I am courteous and affable, good humour'd (unless I am first provok'd), and handsome, and sometimes witty, but always, Sir, Your Friend and Humble Servant,

Silence Dogood

Ben later recalled that his brother was furious (and probably felt a little foolish) when his authorship of the pieces was finally disclosed, as James and his colleagues had spent a good deal of time speculating about which of their acquaintances was submitting them. But James soon had more serious troubles, for the *Courant* writers, while delighting Boston readers with their wit and daring prose, had begun to try the patience of the local authorities. In the issue of June 11, 1722, the Franklins, satirizing a common suspicion that the local admiralty was in cahoots with pirates and smugglers, ran a piece about a cowardly "Captain Peter Papillon" ("Captain Butterfly"), who had vowed to go in pursuit of the pirates, "sometime this month, if wind and weather permit." Terming the article a "high affront," the General Court ordered James arrested, and he and Ben were both interrogated by the court. Ben was considered innocent by virtue of his youth ("they contented themselves with admonishing me, and dismissed me, considering me perhaps as an Apprentice who was bound to keep his Master's Secrets"), but James was imprisoned. "During my Brother's Confinement," Ben later wrote, "which I resented a good deal, notwithstanding our private Differences, I had the Management of the Paper, and I made bold to give our Rulers some Rubs in it, which my Brother took very kindly, while others began to consider me in an unfavourable Light, as a young Genius that had a Turn for Libeling and Satyr."

James was eventually released, but soon again found ways to offend the authorities with his editorials. In February 1723 he was officially denied the right to publish the *Courant*. James briefly considered dodging the edict by changing the name of the paper,

but a friend suggested simply transferring it into Benjamin's name. As Ben was the apprentice, James had to give him his official discharge in order to facilitate this maneuver.

Although he supposedly remained tethered to his brother by a new secret pact of apprenticeship, Ben took the canceling of his original papers to mean he was now free of his obligations, and began to plot his escape from Boston. "I had already made myself a little obnoxious to the governing Party," he wrote, "and it was likely I might if I stay'd soon bring myself into Scrapes, [for] my indiscrete Disputations about Religion began to make me pointed at with Horror by good People." In mid-September 1723, a friend of Ben's told a sloop's captain that young Franklin had to leave town secretly because he "had got a naughty Girl with

Child." The captain of the sloop, which was bound for New York, agreed to look the other way while the runaway slipped aboard. "So I sold some of my books to raise a little money," Franklin recalled, "was taken on board privately, and as we had a fair wind in three days I found my self in New York near 300 miles from home, a boy of but 17, without the least recommendation to or knowledge of any person in the place, and with very little money in my pocket."

The enterprise of being placed in charge of a rebellious newspaper, having his first prose writings praised, and even his unhappy experience with his brother's domineering personality (he accused James of occasionally beating him, "which I took extremely amiss"), contributed to Benjamin Franklin's early self-assuredness and his lifelong distrust of authority. "I fancy his harsh and tyrannical Treatment of me," Ben later said of James, "might be a means of impressing me with that Aversion to arbitrary Power that has stuck me thro' my whole Life." Claude-Anne Lopez, a biographer of Franklin's private life, concludes, "The rebel in him which would one day defy the rule of the Penns, the sovereignty of the king, and the terror of lightning, had taken his initial steps in the brother's printing shop, not amid the father's candles."

Although the *Courant* existed for only five years, it demonstrated that a newspaper could take anti-authoritarian positions and gain a substantial readership. James Franklin was, in a sense, a martyr for early American journalism, for after his release from jail in 1723, Massachusetts never again attempted to censor the press. In a more renowned case a dozen years later, New York printer John Peter Zenger was acquitted of libeling the colonial governor, establishing more broadly the right of newspapers to criticize the government.

Ben Franklin saw Cotton Mather for the last time while on a visit to his family in Boston in the spring of 1724. Mather apparently was willing to excuse his part in the inoculation dispute and may have perceived in the younger Franklin a kindred spirit. "He received me in his library," Franklin recalled, "and on my taking leave showed me a shorter way out of the house through a narrow passage, which was crossed by a beam overhead. We were still talking as I withdrew, he accompanying me behind, when he said hastily: 'Stoop, stoop!' I did not understand him, till I felt my head hit against the beam. He was a man that never missed any occasion of giving instruction, and upon this he said to me: '*You are young, and have the world before you; STOOP as you go through it, and you will miss many hard thumps.*' "

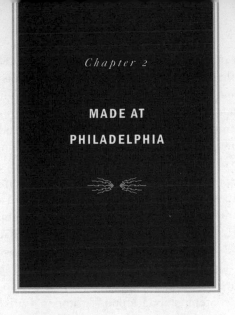

Chapter 2

MADE AT
PHILADELPHIA

PHILADELPHIA, WILLIAM PENN'S CITY ON THE DELAWARE RIVER, WHERE
Franklin's great electrical discoveries would take place, had not
been the runaway apprentice's chosen destination, but it proved a
fortuitous one. Settled later than Boston, Philadelphia drew a more
diverse stream of emigrants from Europe, and—benefiting from
its central location on the eastern seaboard—was by the mid-
eighteenth century the colonies' intellectual and cultural cross-
roads. Britain had lacked the means to tend to all its distant
colonies in America, so royal charters were sold to groups or individuals who in return received vast tracts of land and the authority
to govern. There was supposed to be official oversight of these
ventures, but in practice the proprietors, such as the Calverts of
Maryland or William Penn, were left alone to manage their holdings and attract settlers the best they could. In the case of Penn,
this meant a promise to residents of a large degree of personal freedom and representation.

The first planned community in America, Philadelphia was de-signed by Penn to avoid emphatically the pitfalls of London, which was notorious for its filth and crowded conditions; he in-sisted on wide avenues and ample green spaces to diminish the risk of catastrophic fire and the spread of disease. And Penn's rela-tively generous approach toward the nearby Indians eliminated the expense of constructing vast fortresses and barricades, a policy much in keeping with prevailing local Quaker sentiment.

If seventeenth-century New England had served as the beach-head for Puritanism, eighteenth-century Philadelphia was where America welcomed the Enlightenment. Boston had its starchy ministers and Yankee merchants, but Philadelphia would have its "Franklins"—the many talented scientists, artists, and craftsmen who flourished in the city's tolerant atmosphere. "No colonial city, and few in contemporary Europe," historian Carl Bridenbaugh points out, "could boast the galaxy of intellectual and civic leaders produced by Philadelphia from 1740 to 1776. . . . They it was who . . . naturalized our second great stream of European thought. . . . This process . . . produced upon the banks of the Delaware a city own-ing the first broadly democratic society of modern times."

Franklin, arriving in Philadelphia in fall 1723, a few months shy of his eighteenth birthday, managed to gain a foothold there, as well as a reputation as a competent printer's assistant. He also "made some courtship" to his future wife, Deborah Read. Perhaps inspired by the example of his brother James's formative *wander-jahr,* however, he sailed to London a year later, on November 5, 1724, at the urging of Pennsylvania's governor, Sir William Keith, who had been impressed by Franklin and who offered to help the young man obtain the necessary equipment in London so that he might become the printer for the Pennsylvania government.

In London, as in Philadelphia, Ben impressed employers with his ability as a printer; he also showed off his swimming skills in the Thames, tried his hand at a precocious philosophical tract (which he soon regretted), savored London's cultural attributes as best he could while living frugally, and made some interesting ac-quaintances, including Dr. Hans Sloane, who upon the death of Isaac Newton in 1726 would become president of the Royal Society.

Ben's year and a half in England also revealed to him how easily one can be harmed by the weakness and mendacity of others. Governor Keith had failed to follow through with his promised assistance, and James Ralph, a young poet friend who had accompanied Franklin abroad, behaved badly, and refused to repay a loan. En route home in 1726, perhaps inspired by his disappointment over the conduct of both a mentor and a friend ("Men and melons are hard to know," wise Poor Richard would conclude), Franklin's thoughts turned to the idea of character and how it might be maintained, improved, even perfected. Leaning on Cotton Mather's notion of mixing personal virtue and civic generosity and the idea, born of the Enlightenment, that individuals might be capable of influencing their own destinies, Franklin set out to create his adult self, drawing up a list of personal goals and virtues to which he would adhere.

"I have never fixed a regular design as to life, by which means it has been a confused variety of different scenes," he reflected as his ship carried him back to America. "I am now entering upon a new one; let me therefore make some resolutions, and form some scheme of action, that henceforth I may live in all respects like a rational creature." He vowed to be frugal, honest, industrious, to speak no ill of others, to remain humble, and to try to manage his life in such a way that his own aspirations would complement and mesh with those of his community.

Since childhood Franklin had been a natural leader of his peers; now, after London, where he had won the admiration of workingmen and gentlemen alike for his enterprise, wit, and even his prowess at swimming, he must have begun to feel something of his own exceptional promise, to understand that he possessed a mind and temperament superior to those of other people. At just under six feet, his muscles taut from years of swimming and carrying heavy boxes of type, he also had a commanding physical presence, and his politely inquisitive nature and charm allowed him to mix easily with all classes of people. However, he knew himself to be easily distracted, even reckless. Perhaps his list of exemplary personal traits was an attempt at self-management, a way of supplying for himself the discipline that, in other forms—church, parents, older brother—he had managed to throw off. In this way he turned the Enlightenment motif of empirical observation into a system of self-evaluation. He would make himself into an experiment. He would, above all, attempt to practice complete candor; if recent experience had taught him anything, it was that "Truth and sincerity have a certain distinguishing native luster about them which cannot be perfectly counterfeited."

Certainly one positive influence on his life at this time, though it proved of short duration, was that of Thomas Denham, a Quaker merchant who had met Franklin in London and employed him as a clerk upon their return to Philadelphia. Denham was a worthy role model and gave Franklin his first business experience and education. Having fully reciprocated the older man's friendship, Franklin was heartsick at his sudden death from illness in 1728.

For the next twenty years he adhered as best he could to the life principles he had devised on that voyage home. In 1727 he led the founding of the Junto, a mutual improvement society probably inspired by similar societies of young men created by Cotton Mather. The Junto was a philosophy club crossed with a professional men's association in which members reflected upon and discussed questions of moral, political, or scientific import, and sought to apply the lessons to affairs in their community and their personal lives. Franklin may have attended some of Mather's gatherings as a child in the company of his father, and when he visited Mather in 1724, the minister likely gave Franklin a copy of a newly printed manual he had written about how such groups were to be organized.

Through his successful printing business and, soon, as Philadelphia's leading publisher—of the weekly *Pennsylvania Gazette* and his popular *Poor Richard's Almanac*—Franklin had contact with all sectors of Pennsylvania society. He printed legal and government documents, and ministers' sermons (the bulk of any printing business in colonial times), and through the *Gazette*'s pages, he had the opportunity to shape opinion in his adopted city. It was *Poor Richard's* that succeeded beyond Franklin's own expectations. Popular almanacs could be "a printer's gold mine," for next to the Bible they were often the only book in early American homes, and unlike the Bible, they were reissued annually. *Poor Richard's*, begun in 1732, suited and expressed Franklin's character and interests perfectly—astronomical, astrological, meteorological, and other scientific curiosa mixed with satire, jokes, and cautionary axioms promoting modesty, thrift, and common sense. The almanac was a standard literary form, but Franklin proved more adept at it than his competitors, and as many as ten thousand *Poor Richard's* sold each year, bringing him a steady income. In one of his first attempts at a literary hoax, a form he would employ regularly throughout his life, Franklin drove a competing almanacker named Titan Leeds to distraction by publishing a phony "prediction" of the precise date of Leeds's death. When the victim protested in print that he had not died and called Franklin a "conceited scribbler," Franklin calmly insisted the death must have occurred as he had prophesied because the *real* Titan Leeds "was too well bred to use any man so indecently and scurrilously, and moreover his esteem and affection for me was extraordinary."

The number of Franklin's public roles increased exponentially—founder in 1731 of the Library Company, the nation's first lending library; clerk of the Pennsylvania Assembly and prime mover of the Union Fire Company in 1736; and, in 1737, postmaster of Philadelphia, a position he held until 1753, when he was made deputy postmaster general for the colonies. In 1748 he was elected to the Council of Philadelphia; in 1750 he joined a commission charged with negotiating with the Indians; and in 1751 he became a member of the Pennsylvania Assembly. He was "flattered by all these promotions, for, considering my low beginning, they were great things to me." Aware that they brought an impossible number of new duties in addition to his own work and family, he created written schedules for his activities, closely managing his obligations, work time, leisure, and even his diet.

The core of Franklin's genius seemed to be, at least in part, two related talents—the ability to discern a chain of causation in human affairs, and the knack for devising ways to address problems at their root cause. Thus, in one of his most famous civic improvements, he tackled the problem of keeping Philadelphia's streets clean. Dirty streets were not simply unpleasant, he observed, but the dust and dirt trampled into market stalls by cus-

tomers or blown by the wind were injurious to commerce. On this
basis he sold the shopkeepers on the idea of contributing to keeping the streets around the market swept. One improvement led to
another. "All the Inhabitants of the City were delighted with the
Cleanliness of the Pavement that surrounded the Market, it being
a Convenience to all; and this rais'd a general Desire to have all the
Streets paved; and made the People more willing to submit to a
Tax for that purpose."

Franklin also understood that as important as it was to do good,
it wasn't a bad idea to be *seen* doing good. Never one to shirk physical labor in any case, when he needed to bring fresh paper from
the docks to his shop, he made sure to push his own wheelbarrow
up Market Street, convinced there would be a double value in honest work performed under the public gaze. Virtue alone was fine,
but how much more would it be worth if joined with the respect of
one's peers?

> In order to secure my credit and character as a tradesman, I
> took care not only to be in reality industrious and frugal, but
> to avoid all appearances of the contrary. I dressed plain and
> was seen at no places of idle diversion. I never went out a
> fishing or shooting; a book, indeed, sometimes debauched
> me from my work, but that was seldom, snug, and gave no
> scandal . . . Thus being esteemed an industrious, thriving
> young man, and paying duly for what I bought . . . I went on
> swimmingly.

Franklin's efforts at self-improvement did not fail to make an
impression. Remarked a neighbor, Dr. Baird, "The industry of that
Franklin is superior to anything I ever saw of the kind: I see him
still at work when I go home from Club, and he is at Work again before his Neighbors are out of bed." James Logan, William Penn's
former secretary and the town's resident learned man, had been impressed that, in addition to his business and civic affairs, Franklin,
while sitting at his desk as clerk of the Pennsylvania Assembly, had
managed to concoct magical number squares, a mathematics game,
superior to those published in a well-known book on the subject.

"Our Benjamin Franklin is certainly an Extraordinary Man in most respects," Logan noted, "one of a singular good Judgment, but of Equal Modesty."

On his return voyage to the colonies in 1726, when Franklin turned his pen to the aspirations he held for his character, he also for the first time gave serious expression to his scientific curiosity. In a shipboard journal, he carefully jotted his observations of rainbows, dolphins, seaweed, crabs, and flying fish, and gauged the remaining distance to Philadelphia by measuring an eclipse of the moon. Already apparent in this writing—he was only twenty—is Franklin's gift for fusing his own cautious hypotheses with what he knew of established science.

Back home, he took delight in Philadelphia's small but growing number of astronomers, naturalists, artists, and literary men. Science, or natural philosophy, as it was known, was particularly fashionable. "An interest in science was considered one of the distinguishing characteristics of a gentleman," a historian of the period has noted, "almost as necessary as foreign travel and the ability to dance proficiently." Prominent in Franklin's circle were John Bartram, a botanist who traipsed the eastern seaboard discovering America's native plant and animal specimens; James Logan, known as the first person in the New World to own a copy of Newton's *Principia;* artist and cartographer Lewis Evans; and, later, astronomer David Rittenhouse. Through an ever expanding web of intercolonial correspondence, Franklin made the acquaintance of the minister Ezra Stiles, the president of Yale College; John Winthrop, the Hollis Professor of Mathematics and Natural Philosophy at Harvard; New York's lieutenant governor (and natural philosopher), Cadwallader Colden; Alexander Garden and the Charleston, South Carolina, botanists; from Virginia, physician Dr. John Mitchell; and, later, Thomas Jefferson, who, along with George Washington, bought cuttings from Bartram's garden on a visit to Philadelphia.

Helping to spark the interest of Franklin and other Americans in natural philosophy was Europe's eagerness to know the flora and fauna of the New World. America itself—thirteen sparsely populated provinces on the shore of an unknown continent—was viewed by European men of science as a vast Arcadian "laboratory" of new species of bird, plant, and animal life, of strange geologic features, and of mysterious native peoples. This fascination found expression in Europe in the advent of *the garden,* a place where the exotic splendor of diverse species could be collected and studied with the help of a recent classification system introduced by the Swedish botanist and physician Carl von Linné (known as Linnaeus). The system used binomial nomenclature, naming items by genus, then by species—such as *Agapanthus africanus* or *Spathodea campalata*—a system that was soon adopted worldwide, as Linnaeus ultimately identified twelve thousand species, separated by their sexual pat-

terns into twenty-four categories. Beginning in 1749, the director of France's Jardin du Roi, Georges-Louis Leclerc de Buffon, produced a forty-four-volume *Histoire Naturelle* that popularized Linnaeus's efforts, giving the world the first accessible ordering of botanical and zoological terms.

Astronomy, with its elucidation of long-mysterious celestial phenomena, had dominated the seventeenth century; botany, and the Linnaean passion for hunting down, identifying, and classifying plant species, became the fixation of the eighteenth. "The closer we get to knowing the creatures around us," said Linnaeus, "the clearer is the understanding we obtain of the chain of nature, and its harmony and system, according to which all things appear to have been created." Of equal importance, in an age when many

physical maladies remained hopelessly beyond cure or comprehension, the cultivation of plants held the potential means of alle-
viating suffering. "We repair our bodies by the drugs from
America," Joseph Addison had acknowledged.

The major gardens of Europe—Kew Gardens in England, the
University of Leyden in the Netherlands, "Physick Garden" at
Oxford, Buffon's Jardin du Roi in Paris, and the many other large
gardens owned by the nobility—relied strongly on imported seeds
and plantings, and vied with one another for exotica. Britain's Lord
Petre boasted a garden of ten thousand American plants, while the
London Quaker physician John Fothergill had thirty-four hundred
species from around the world. One European scholar planted part
of his estate in Thuringia so densely with New World seedlings
that he named it "America." Philadelphia's John Bartram, appointed botanist to the king by George III, is thought to have sin-
gle-handedly introduced more than one hundred North American
plant species into Britain. Even Franklin, who did not botanize to
any great extent, got into the act, sending the first Venus flytrap
across the Atlantic to take root in Buffon's flower beds.

Linnaeus, Bartram, and many others were aided in their transatlantic exchanges by Peter Collinson, a London wool merchant who specialized in the importation of rare plant and animal species, and who serviced many of the English estate gardens. Collinson, who supplied books to Franklin's Library Company and who would be the initial sponsor of Franklin's electrical ideas in England, was descended from the earliest English Quakers, who promoted gardening as an aesthetically satisfying alternative to entertainments forbidden to their sect, such as the theater. He had been admitted to the Royal Society in 1728, although it appears his only direct contribution to science was his successful refutation of the notion, held even by Linnaeus, that swallows hibernate underwater. "A knowledge of science was the passport to the intellectual society of the 18th century," Bridenbaugh writes, "and it was Peter Collinson of London who issued the visas to the initiate of many lands."

"Keep the chain bright and shiny," the Native American phrase that Collinson, Franklin, and their cohorts employed to describe their efforts at international scientific cordiality, was entirely appropriate: Durable linkages were what they hoped to fashion. In an

age when journals and newspapers appeared irregularly, and cir-
cumspection was requisite in airing one's views, correspondence
was where the century's ideas truly flowered, and where some of
its most fervent passions were expressed. Bartram, writing to
Franklin in England in 1757, urged his friend to "keep the chain of
friendship bright while thee art diverting thy self with the generous conversation of our worthy friends in Europe," and begged a
few lines of Franklin's correspondence, for "they have A Magical
power of dispelling melancholy fumes."

Bartram was the prototype of the American backwoods naturalist, covering hundreds of miles on foot or horseback, evading potentially hostile Indians, and poking for specimens amid rocks and
on steep mountainsides. He had, through his friendship with
Collinson, published eight papers in the *Philosophical Transactions*,
mostly regarding his two favorite subjects, wasps and snakes, the
latter of which he was known to capture with his bare hands.
Surmising that there must be "very great Forests and a fertile
Country to the Westward," Bartram in 1743 made a memorable
journey into western Pennsylvania and then up along the shore of
Lake Ontario in the company of the Philadelphia mapmaker and
artist Lewis Evans, and Conrad Weiser, who acted as ambassador
to the Indians for the colonies. Bartram's journal of this trip, published in London in 1751, brought Linnaeus's appraisal that he was
"the greatest natural botanist in the world."

Franklin, as postmaster, could offer Bartram and other American
botanists invaluable aid with the shipment of their precious specimens to England. Apples from America were always in great demand, as were such exotic souvenirs as Indian bows and arrows,
turtles, melons, and animal hides. The traffic involving plants was
the most risky, and not infrequently a British patron opened an arriving crate only to find eagerly awaited trees or shrubs in a de-
plorable state. Many died of cold aboard ship, were chewed on by
rats, or were overwatered by ships' captains during the four-to-
eight-week transatlantic passage. Franklin's friend Dr. John
Mitchell, sailing to England in 1746 with one thousand plant specimens, had the worst hard-luck story; mid-ocean, his ship was
boarded by privateers, his collection was ruined, and even his per-

sonal property was stolen; he arrived in London with only the clothes on his back. Collinson, the middle man in countless transatlantic arrangements, heard almost continual demands and complaints from his contacts on both sides of the water. Bartram always wanted more money, while the gentry Collinson assisted in England were quick to demand he make house calls when valuable imported plants refused to thrive. Franklin, who had protested England's practice of transporting its felons to America, took revenge when given the chance to ship live rattlesnakes across the water. "Rattlesnakes seem the most suitable returns for the human serpents sent us by our Mother Country," he wrote, noting that at least "the Rattlesnake gives warning before he attempts his mischief, which the convict does not."

Franklin's wide acquaintance with men of scientific bent, and his general optimism about the growth of interest in natural philosophy, coalesced in 1743 in his suggestion for a Philadelphia-based scientific organization. In his "Proposal for Promoting Useful Knowledge Among the British Plantations in America," co-authored with Bartram, Franklin called for "One Society [to] be formed of Virtuosi or ingenious Men residing in the several Colonies, to be called The American Philosophical Society; who are to maintain a constant correspondence," and whose attention would be given to:

All new-discovered Plants, Herbs, Trees, Roots, their Virtues, Uses, &c.; Methods of Propagating them, and making such as are useful, but particular to some Plantations, more general. Improvements of vegetable Juices, as Cyders, Wines, &c.; New Methods of Curing or Preventing Diseases. All new-discovered Fossils in different Countries, as Mines, Minerals, Quarries, &c.; New and useful Improvements in any Branch of Mathematicks; New Discoveries in Chemistry, such as Improvements in Distillation, Brewing, Assaying of ores; &c.; New Mechanical Inventions for Saving labour; as Mills and Carriages, &c.; All new Arts, Trades, and Manufactures, &c. that may be proposed or thought of; Surveys, Maps and Charts of particular Parts of the Sea-coasts, or Inland Countries; Course and Junction of Rivers and great Roads,

Situation of Lakes and Mountains, Nature of the Soil and
Productions; &c. New Methods of Improving the Breed of
useful Animals; Introducing other Sorts from foreign countries. New Improvements in Planting, Gardening, Clearing
Land, &c . . .

Despite initial enthusiasm from Franklin's counterparts for this
ambitious project, and his own high hopes, the Society faltered, as
no one other than Franklin seemed sufficiently dedicated. Although
he remained confident in the burgeoning scientific collegiality in
the colonies, a generation would pass before the Society was successfully revived.

Franklin, while knowledgeable about botany and willing to
abet its development as best he could, never immersed himself as
deeply in the subject as did many of his friends; one suspects he
was simply too busy, or of the wrong temperament, to devote long
hours to hunting specimens. His first noteworthy contribution to
colonial science came in a different field in about 1739 or 1740,
with the design of his "Pennsylvania fireplace."

Colonists used fires for heat and for cooking, but never with
much thought as to how a fire's heat might be more effectively dis-
tributed and conserved. With traditional open fireplaces, a great
deal of heat was lost, a problem Franklin believed could be addressed if the smoke that ordinarily ascended upward through the
chimney could be made to first "descend" before it was allowed to
exit. His initial interest in the subject was probably stirred by
Pennsylvania iron manufacturer Robert Grace, one of the members of his Junto, who gave Franklin an English translation of *La
Mécanique du Feu*, the 1713 tract by Nicolas Gauger in which the
author describes a fireplace with hollow iron jambs and an air
pocket that draws in cold air and circulates warm air heated by the
fire.

In Franklin's design, "air was drawn in through a duct at the
bottom of the stove, circulated through the air-box around a series
of baffles, and passed into the room through openings on both
sides of the stove near the top." The fire's warmth, instead of rising immediately upward and away from the room, would wrap itself through a chamber in the stove, radiating heat outward into the

room. He also brought the stove away from the exterior wall of the house and placed it toward the center of the room, so that occupants could benefit from the heat given off by all the stove's surfaces. Finally, its movable front doors gave the user the choice of whether or not to enjoy the sight of the fire itself, an option denied to them by closed stoves. People not only felt comforted by the sight of a fire in the hearth, Franklin knew, but were made nervous by a "fyre secret felt but not seene." His stove thus allowed for the coziness of the fire but alleviated the need for people to huddle directly in front of it, "where many catch cold," Franklin observed, "being scorcht before, and, as it were, froze behind."

In 1742 Franklin gave Grace a model of his stove, for which the ironmaster then cast plates at his furnace. As he would with all his inventions, Franklin, although he stood to profit from sales of the stove, did not apply for a patent. He believed that products of the human imagination belonged to no one person, and should be shared by all. Besides its commendable altruism, this philosophy probably saved him from a tremendous amount of aggravation. Anyone seeking to patent a new mechanical innovation in the New World would need to secure it in each colony individually, as until 1790 there was no national patent system in America. He was also spared the nasty proprietary disputes and jealousies that often embittered other inventors, the kind of contention he abhorred.

Ultimately, the design of his stoves may have been a bit overambitious; in practice they didn't work as efficiently as he had hoped, and customers often wound up modifying them for everyday use. The "Franklin stove" of today is related only vaguely to the original design. However, what was most significant about Franklin's invention was that a potentially useful product had resulted from his resourceful application of pure scientific theory. This example he would soon repeat to even more stunning effect.

It is not entirely clear what first set Franklin on a course of electrical experimentation, although the appeal of electricity to Franklin (or to anyone fascinated by storms, ocean currents, and movements of warm air in a room) is readily understandable: Electricity, like gravity or heat, appeared to be a "subtle fluid," a force with physical properties but no mass. What is known is that on a visit to his family in Boston in late spring 1743, Franklin learned of a proposed lecture series in "Experimental Philosophy" to be given by Dr. Archibald Spencer of Edinburgh, a male midwife and itinerant lecturer. Spencer was to talk on diseases of the eye and physiology, and demonstrate some electrical exhibits. The requisite number of participants was never attained, so the course was not given, although Franklin seems to have befriended Spencer, for as Franklin remembered, he "show'd me some electric Experiments. . . . They were imperfectly performed, as he was not very expert; but being on a Subject quite new to me, they equally surpriz'd and pleas'd me." A year later, Franklin encountered Spencer again when the lecturer passed through Philadelphia on his way to the South and the Caribbean. Spencer offered to distribute copies of Franklin's "Proposal for Promoting Useful Knowledge" in the West Indies. Franklin in turn agreed to serve as an agent for a series of Spencer's talks in Philadelphia, the *Gazette* instructing interested parties to come to Franklin's post office on Market Street to subscribe and pick up a catalog.

In Britain, electrical demonstrations of the kind Spencer gave were sometimes called "electricity parties," for scientific lectures generally were seen as entertainments. They were often based around mechanical wonders, such as life-size models of the human torso that displayed the workings of the heart and circulation system, musical clocks with moving figurines that performed well-loved tunes by Handel and Corelli, and the camera obscura that projected images of celebrated battle scenes against the wall. A typical ad for a lecture that Franklin ran in the *Gazette* in 1744 boasted of a magnifying solar microscope that would reveal:

The Animalculae in several Sorts of Fluids, with many other living and dead Objects, too tedious to mention . . . shewn most incredibly magnified, at the same Time distinct; also the Circulation of the Blood in a Frog's Foot, a Fish's Tail, also in a Flea, and Louse, where you discover the Pulse of the Heart, the moving of the Bowels, the Veins and Arteries, and many small Insects, that one Thousand of them will not exceed the Bigness of a Grain of Sand, with their Young in them; [and] Eels in Paste, which have given a general Satisfaction to all that ever saw them.

No copy of Spencer's catalog survives, but written accounts of his Philadelphia lectures, which were given in the Library Company's rooms in the Pennsylvania State House (later, Independence Hall), indicate that many notables attended and that Spencer, a "most judicious and experienced Physician and Man-Midwife," discussed the workings of the eye, Newton's ideas about light and color, and William Harvey's theories of the human circulation system. Spencer mixed pure science with an ingratiating down-home wisdom, reminding his listeners that in case of fever, "A little Cool Air is proper . . . but no Cordials."

The most novel of Spencer's demonstrations were the electrical ones, which employed a glass tube, a piece of sheepskin, a cork ball, and other accessories. In a series of "experiments," Spencer sent mild shocks through volunteers, caused loud sparks to explode, and used static electricity to make objects such as paper bits and metallic fragments move as if by magic. When he took a long glass tube, rubbed it, and then held it near brass or gold leaf, it "put them into very brisk & Surprizing Motions." Franklin, intrigued, arranged to buy Spencer's electrical demonstration equipment when his lecture tour ended.

At the time, electricity was a field of natural philosophy wide open to the curious layman, and thus an ideal subject for someone of Franklin's background. Unlike astronomy, medicine, or botany, which required specialized knowledge, electricity could be understood—and at this point was more or less *only* understood—as a plainly observable attraction between objects. Franklin, who was familiar with the *Philosophical Transactions* and other British magazines and books on scientific subjects, had likely read articles about the present state of electrical knowledge by William Watson, England's most prolific electrical experimenter. Watson had built on the work performed originally in Germany by Georg Mathias

Bose, professor at the University of Leipzig, one of electricity's first great showmen, who created "electric halos" over soldiers' heads and "electric kisses" so powerful that men who kissed Bose's "electrified girls" occasionally cracked their teeth. Bose had recently shown that warmed spirits could be ignited with an electric spark, and had worked on the electrical firing of various other substances, including gunpowder.

It is doubtful that Franklin could have followed closely Newton's monumental *Principia* (few could), but he certainly understood its conclusions, and he was a committed Newtonian for his faith in experimentation. He was also well-acquainted with the other advances of the Scientific Revolution—William Gilbert's treatise on magnetism, William Harvey's explanation of human physiology, Edmond Halley's predictions about comets—and the new scientific tools such as the thermometer, the microscope, and the telescope. (He had tried, unsuccessfully, to meet Newton, then the president of the Royal Society, during his visit to England in the 1720s.) But Franklin was chiefly taken with another of Newton's works, the more straightforward *Opticks* (1718), a treatise on light phenomena and color that offered experimentation, observations, and the finding of proofs—induction as opposed to deduction—as a means of reaching scientific conclusions. "In the *Opticks*," notes I. Bernard Cohen, "Franklin and other scientists of his generation found a veritable handbook of the experimental art." Many knew by heart the book's creed, which appeared in a final paragraph:

As in mathematics so in natural philosophy, the investigation of difficult things by the method of analysis, ought ever to precede the method of composition. This analysis consists in making experiments and observations, and in drawing general conclusions from them by induction, and admitting of no objections against the conclusions, but such as are taken from experiments, or other certain truths.

When Franklin began his work on electricity in the 1740s, most of what was known on the subject had been gained only recently. Although humans had observed the effects of static electricity since before the time of Christ, until about A.D. 1600, all that was understood was that amber, the translucent, fossilized resin of prehistoric trees, had a mysterious attractive power toward other, lighter bodies. Amber itself had long been the subject of myth. Phaeton, it was told in Greek mythology, the child of the Sun, stole his father's chariot and drove it recklessly through the heavens. Jupiter, watching with alarm, dispatched a thunderbolt that sent Phaeton crashing to earth, where he drowned in a river. His sisters, the Heliades, stood so long weeping by the river's banks that the gods changed them into poplars, and the golden tears they shed became glassy amber. The Greeks thus looked upon amber "with superstitious reverence, and even thought that it had a soul. For when it was rubbed, it seemed to live, and to exercise an attraction upon other things distant from it." Amber's Greek name, *elektrum*, or *elektron*, is derived from the word *elektor*, meaning "sun" or "shining one." Amber, in other words, was bottled-up sunshine.

First-century philosopher Pliny the Elder, whose *Natural History* was probably the world's first science bestseller, wrote of amber's electrostatic properties, and also observed that there was a fish, the *torpedo*, that gave off electrical shocks when handled. He passed down a story of men who had been cured of the gout by touching this fish, no doubt history's original mention of electricity as a medical treatment. Before Pliny's scientific curiosity could guide him to more revelations about electricity, however, it led him in A.D. 79 to Pompeii, where he became one of the victims of the eruption of Mount Vesuvius. After Pliny, the subject of electrostatics virtually disappears for the next fifteen hundred years.

Renewed efforts toward understanding the forces of electricity arrived in 1600 with the publication of William Gilbert's monumental treatise on magnetism, *De Magnete*. Gilbert, physician to Queen Elizabeth, demonstrated for the queen and her court the principles of terrestrial magnetism with a "terella," a model of the earth made of lodestone. As a sidebar to his primary investigation, he looked at attractions that were not magnetic, and found that substances like glass and amber, when rubbed, attracted lighter objects such as feathers. Gilbert gave the phenomenon the name "electric," from the Greek term for amber, and referred to substances that showed these properties as "electrica" (the word "electricity" would not be used until 1646). However, his main topic being magnetism, he devoted only a single chapter to "electrica" in his six-volume treatise. ("Gilbert may justly be called the father of modern electricity," a later history would acknowledge, "though it be true that he left his child in its very infancy.")

The next step forward in electrostatics came when investigators, weary of the vigorous rubbing required to generate sparks, created more powerful friction devices, such as that made in 1671 by Otto von Guericke, mayor of Magdeburg, Germany. Von Guericke was already famous for his "Magdeburg hemispheres," two copper hemispheres fitted together that, when the air was sucked from between them by a vacuum pump, could not be separated even by two teams of horses. To obtain consistent electrical charges, he placed a sulfur sphere six inches in diameter into a wooden frame and rubbed it with one hand while turning the sphere with an iron handle. Taking the charged sphere from its frame, Von Guericke used it to chase a feather around the room. He found that it also glowed in the dark and, if held to his ear, emitted "roarings and crashings." Related work was done in England in 1709 by Francis Hauksbee. Hauksbee manufactured a glass globe, eight or ten inches in diameter, then inserted it in a frame, and turned it swiftly on its axis as a means of generating large sparks. He discovered that holding the charged globe near one's face created a mysterious energy he called the "electric wind."

One of the most important early experimenters was Stephen Gray, a dyer by trade who, having suffered a workplace injury, lived as a pensioner in the Charterhouse School, a London private school that also functioned as a retreat for older gentlemen. In 1732 he became the first to show that some bodies can conduct electricity and others cannot, and that the power of electrical attraction, what he called "the electric virtue," could be transmitted. In one experiment he placed a cork in the end of a hollow glass tube, inserted a stick into the cork, and then affixed an ivory ball at the opposite end of the stick. Rubbing the glass tube, he found, electrified the ivory ball. In further experiments carried out at the urging of Granville Wheeler, a wealthy member of the Royal Society, Gray showed that an electrical charge could be sent through packthread, a firm two-ply twine. An early test sent a charge 147 feet; another traveled an astounding 765 feet. Eventually he found he could send a charge a distance along the thread without even touching the line to the tube, but simply by bringing it nearby—proof that no direct connection was needed to transmit a charge, that electricity could be drawn freely through air toward a nearby contact.

Gray was the first investigator to distinguish between conductors, substances such as metal and water through which electricity is freely transmitted, and nonconductors, such as wood, where the charge remains fixed only near the point of contact. Assisted by Wheeler and one of Wheeler's servants, Gray began methodically electrifying any object he could lay his hands on—an umbrella, a live chicken, a teapot, a hot poker—in an effort to ascertain their electrical properties. He discovered that a conductor cannot be electrified while it touches the earth because the charge is immediately carried away to the ground. A conductor supported by a nonconductor, in order to keep it from touching the earth, was deemed to be "insulated." However, Gray found that "if an insulated charged conductor is connected to the earth by any conductor, such as a wire, the charge immediately disappears"—one of the concepts that Franklin would eventually make integral to the lightning rod. Gray was also among the first people to remark that the aural and visual effects of electrical sparks resembled thunder and lightning.

Aided by his unique living arrangements at the Charterhouse, where he was surrounded by boys of various ages and sizes, Gray soon conceived and executed the experiment that would make him famous: the electrification of a human being. Suspending a small volunteer from the ceiling of his laboratory with multiple cords of silk, thus insulating him, Gray electrified a glass tube with a friction machine, then with the glass tube made the youngster's hair stand on end, and drew sparks from his nose and fingers. This dramatic demonstration, known as "The Dangling Boy," became one of the most admired parlor tricks of the age. To heighten its appeal, experimenters often asked one of the more dignified or conspicuous members of the audience—a bewigged magistrate or an elegantly dressed young belle—to come forward and draw sparks from the child's extremities.

Across the English Channel, Gray had a counterpart in French naturalist Charles-François de Cisternay Du Fay, who repeated some of Gray's experiments but bettered his transmission of electricity by sending a charge 1,256 feet over a wet string. Du Fay also built on Gray's work by categorizing those objects that were capable of becoming electrified, those that were not, and the powers of attraction and repulsion between substances. He found that objects that repelled one another possessed the same kind of electricity, that those containing opposite types would attract, and that certain materials functioned as insulators from any electrical charge. Empowered by his observations, in 1733 Du Fay offered the first "complete" theory of electricity, naming the two kinds of electricity—*vitreous*, as found in glass, wool, and hair, and *resinous*, as exists in amber, silk, thread, and paper.

Many improvements were made in the efficiency of "electrical machines" in the 1730s and 1740s. Leather cushions replaced the use of experimenters' hands in generating friction, and the glass globe introduced by Von Guericke and Hauksbee was swapped in various models for a glass cylinder or a glass plate, while a foot treadle replaced the hand crank. These labor-saving devices facilitated the role of nonscientists in electrical experiments, often turning them into entertainments, and helped create the increased reportage on electrical phenomena that piqued Franklin's curiosity.

Late 1746 and early 1747, when Franklin began his work, proved a particularly fortuitous time for taking up electrical experiments. Physicist Pieter van Musschenbroek, at the University of Leyden, had recently perfected the first method in which electricity could be captured, stored, and moved from one location to another. The condenser, or Leyden jar, as it was called after the city of its origin, was a glass container with iron filings at the bottom, filled with water, its walls coated inside and out with tinfoil, with a wire through the top leading down into the water. It greatly improved an investigator's ability to store electricity for use in experiments, well beyond that available from friction machines. In fact, the spark that could be obtained from a Leyden jar was astonishingly powerful, so much so that it nearly killed its inventor. Musschenbroek, naïvely touching the wire leading from the prime conductor, received so large a shock as to be instantly convinced he had become the young science's first martyr. Writing of his discovery to René A. F. de Réaumur, his friend at a scientific academy in Paris, Musschenbroek referred to his breakthrough as a "new but terrible experiment, which I advise you never to try yourself, nor would I, who experienced it and survived by the grace of God, do it again for all the kingdom of France."

The capability of the Leyden jar was soon being explored thoroughly by the Abbé Jean-Antoine Nollet, France's leading electrical scientist and a protégé of Du Fay. Nollet was the author of the provocative theory that electricity was an effluvium, one that flowed in and out of objects' "pores," a theory that countered the long-held belief that electricity was a kind of fire. A welcome figure at Versailles—he was tutor to the dauphin—Nollet delighted the court of Louis XV with a unique demonstration of the Leyden jar's power. He had 180 soldiers gather in a circle and grasp hands, and then sent them simultaneously leaping into the air at the application of a single shock. On another occasion he performed with the help of two hundred Carthusian monks, who held small iron rods between their hands in a long semicircle formed before the king. So talked about were these feats, later known as "circle charges," that other electricians were soon assembling great lines of volunteers—in the Tuileries, across rivers in Germany and canals in Holland, and in the public squares of St. Petersburg. In London, William Watson used Westminster Bridge for a dramatic experiment to find out how far an electrical shock could be transmitted instantaneously, sending a charge across the Thames for a distance of 12,276 feet, more than two miles.

Franklin's transatlantic friend Peter Collinson, the London merchant so encouraging to colonial science, watched such events with interest. He suspected that the snaps and sparks of static electricity were the observable manifestations of a seminal force in nature, one whose mysteries Newton might well have solved had he lived longer. Knowing of his American friend's fondness for making experiments, Collinson sent Franklin a glass tube similar to the

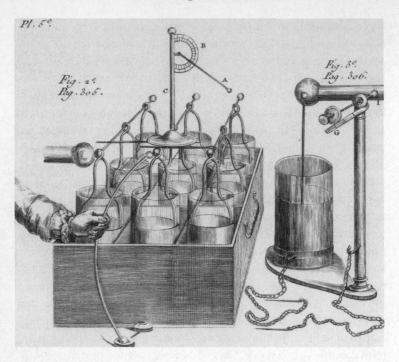

one Europeans had been using in their work on electrostatics. Also included in the package was an article on the present state of electrical knowledge that had appeared in the April 1745 edition of *Gentleman's Magazine*, an English review of current events. Written by Albrecht von Haller, a Swiss professor at the University of Gottingen in Holland, it retraced much of what was known about electricity, and described many of the experiments Franklin would build on, such as igniting rum or other spirits with an electric spark, and the startling and sometimes painful "electric kiss."

"Electricity is a vast country, of which we know only some bordering provinces," the Von Haller article concluded. "It is yet unreasonable to give a map of it, and pretend to assign the laws by which it is governed." Franklin, closing the magazine, may have considered how such a remark could be applied to the American colonies themselves. "The first Drudgery of Settling new Colonies, which confines the Attention of People to mere Necessaries, is now pretty well over," he had written in 1743, "and there are many in every Province in Circumstances that set them at Ease, and afford Leisure to cultivate the finer Arts, and improve the common Stock of Knowledge." Now, with Collinson's letters and the Von Haller article in hand, Franklin had decided to try and live this latter ideal, freeing himself from the daily demands of his business in order to indulge his love of science. He arranged for David Hall, a man recommended by Franklin's London friends, to come from Edinburgh to take over the operation of his printing business.

In comparison with other colonial families, in which the number of children often exceeded ten, and in which one husband might wear out two or three wives, Ben and Deborah had a small, manageable family—Franklin's son, William, born in 1731 to a woman other than Deborah but raised by her and Franklin, who was now, in his father's esteem, "a tall proper Youth and much of a Beau," and Sarah, or "Sally," born in 1743. This meant the Franklins lived with modest financial demands, given the sound management of his business and the post office and stationery shop they ran out of their house on Market Street.

"I am in a fair Way of having no other Tasks than such as I shall like to give my Self, and of enjoying what I look upon as a great Happiness, Leisure to read, make Experiments, and converse at large with such ingenious and worthy Men as are pleas'd to honour me with their Acquaintance," Franklin wrote to Cadwallader Colden, warmly anticipating the new arrangement's potential.

Soon he was writing excitedly to Collinson, "I never was before engaged in any study that so totally engrossed my attention and my time as this has lately done, for what with making experiments when I can be alone, and repeating them to my friends and acquaintances, who . . . come continually in crowds to see them, I have, during some months past, had little leisure for anything else." Franklin quickly became skilled at giving electrical demonstrations. "My House was continually full for some time, with People who came to see these new Wonders." Soon tiring of the repetition, however, he ordered several glass tubes of different sizes made, based on the one Collinson had sent, and taught his friends how to draw sparks, so that "we had at length several Performers." A story that Franklin, seeing some neighbors peering in at his efforts through an iron fence, sent an electric shock through the fence, startling them, is most likely apocryphal, but suggests the local notoriety that attended this phase of his work.

Three of Franklin's friends from the Junto—Thomas Hopkinson, a lawyer and the first president of the fledgling American Philosophical Society; Philip Syng, a silversmith whose inkstand would be used at the signing of the Declaration of Independence; and Ebenezer Kinnersley, a Baptist clergyman temporarily without a pulpit—joined him in a more focused series of experiments. One of the first demonstrated how electricity is not "manufactured" by friction, but simply redistributed. A, by rubbing a charged globe, electrified it; B then received a charge from the globe; when C then touched knuckles with B, C got a shock because B, having received additional electricity, was positively charged; if C then touched A, the latter received a shock, because A, having given his electricity to the globe, was negatively charged. The only way shocks would not be given, Franklin saw, was when both parties touching were equally charged. Franklin's efforts to explain these attractions and repulsions led him ultimately to conclude that electricity doesn't vanish but only moves from object to object and is always present in some amount. This theory—the conservation of charge—represented a key breakthrough, as it helped lead Franklin to a means of characterizing the changes observed in electrified objects.

Electrical studies had up to this point not centered on the idea that electricity was a substance. Rather, it was considered a kind of power or "virtue" that could be excited in bodies. It manifested itself in the power to attract nearby light bodies and, later, the ability to deliver sparks. Gray and Du Fay had introduced the concept that the virtue could be transferred between bodies. But what was behind this virtue? How could it be explained?

Franklin and other electrical investigators suspected that all ordinary matter must be suffused with some active subtle matter that bodies acquired and that caused electrical phenomena. Some thought of it as a kind of "aether" that literally filled not only all matter but all space. Others thought it could exist only in ordinary matter. These "first primitive views," as historian I. Bernard Cohen terms them, "that electricity is due to 'something' in bodies which could change in amount or in distribution," were sharpened by Franklin and Nollet, who showed the way to the notion that electricity was a matter present in *all* bodies, a distinct kind of matter that is always the same. They may have been better able to make this leap than their predecessors because of their extensive work with the newly invented Leyden jar. Because the jar could store electricity, it allowed them to move electricity about as part of a graduated process, not merely to see it as the sudden flash that occurred between objects in a friction experiment.

Nollet built on the theory advanced by his mentor, Du Fay, of two electrical fluids, one resinous, the other vitreous, and hypothesized that electricity resulted from two simultaneous streams of effluent and affluent fluid that were set into motion when an electrifiable body was rubbed. Franklin differed, suggesting that the subtle matter of electricity was better defined by its accumulation or deficit in electrifiable bodies. While Nollet's description of effluence and affluence dealt with varying *quantities* of electrical matter that he supposed excited one another and flowed simultaneously in opposite directions in the vicinity of electrified bodies, Franklin's theory gave a more comprehensive account of two *qualitatively* different types of a single fluid that displaced each other.

"Perhaps reflecting his business experience," suggests historian Patricia Fara, "Franklin's fluid behaved rather like money: just as a bank account may be in credit or overdrawn, an object may have more or less than its normal equilibrium amount of electricity, thus making it appear to be positively or negatively charged." Improving on Du Fay's "vitreous" and "resinous" designations of supposedly "opposing" electricities, Franklin introduced the more accurate terms "positive" and "negative" to describe the differing qualities of surplus or deficit. He also confirmed Stephen Gray's suggestion that bodies involved in an electrical exchange need not touch—that electricity, like magnetism, could exert its power at a distance.

Franklin's theory of electrical action was elegant in its simplicity. Electricity was not "fire"—although he would continue to use that term euphemistically—and not an effluvium, but another kind of elemental force in nature, one contained in all objects, which become electrified by gaining or losing their share of it.

"The electrical matter consists of particles extremely subtle, since it can permeate common matter, even the densest, with such freedom and ease as not to receive any appreciable resistance," Franklin wrote. This was another significant insight, an atomic theory of electricity, one that would not be vindicated for another century and a half, until 1897, when British physicist Joseph John Thomson discovered the electron. The speculation that matter comprised atomic particles too small for the eye to see had existed as far back as Aristotle, but of course no one, not even Franklin, imagined that the corpuscles that caused electrical phenomena were smaller even than an atom, or that it would one day be possible to observe these "invisible" particles.

His focus on how electricity accumulates in and may then be withdrawn from an object was what initially set Franklin on the path to the lightning rod. He reported to Collinson about his experiments with the points of iron bodkins, which Franklin found would draw off an electrical discharge. "Since all bodies have a 'normal' quantity of electric matter, as much as they can hold," Cohen explains, "any excess that they receive must be outside their exterior surface or collect around that surface to form a loosely bound 'electrical atmosphere.' Hence, if a positively charged body were brought near a negatively charged body, there would be an attractive force between the unsaturated common matter in the negatively charged body and the excess electrical matter bound to the positively charged body." Franklin's colleague Thomas Hopkinson was instrumental in leading Franklin and the others to understand this power. When the experimenters were working to electrify shot in the mouth of a bottle, Hopkinson suggested that a more powerful spark could be had by laying a needle across the shot. When this was done, however, it proved impossible to electrify the shot because the point of the needle drew off the electrical charge. "If you present the point [of a long needle connected to earth] in the dark," Franklin wrote to Collinson, "you will see sometimes at a foot distance or more, a light gather upon it like that of a fire-fly or glow-worm; the less sharp the point the nearer you must bring it to observe the light; and at whatever distance you see the light, you may draw off the electric fire."

Franklin concluded that a sharper iron point is more effective in drawing off an electrical discharge than a blunt one, a "law" he would wind up defending vigorously in later controversies over the utility of the lightning rod. To illustrate the principle, Franklin compared the relative powers of a blunt object and a pointed one to someone removing hairs from a horse's tail. Trying to grab a handful of the horse's hair won't work as well as using your fingers to pluck one hair at a time. In the same way, a pointed rod will more easily draw electrical particles.

Franklin and the other Philadelphia experimenters were not all seriousness. Like their British counterparts who staged "electricity parties," Franklin and his associates developed numerous electrical novelties with the help of the substantial charge attainable with a Leyden jar. They made bells ring and wires jump, ignited barrels of rum, devised "toys" such as an animated "electrical spider," and staged tricks like "invisible candle-lighting" by electricity. Women visiting Franklin's workshop were cajoled into serving as "electric Venuses" who then planted "electric kisses" on the cheeks of male volunteers. He described conducting "circle charges" with groups of friends, and a demonstration he named "The Conspirators," in which he made sparks flicker about the edge of a gilt-framed portrait of the king. Franklin also informed Collinson that he had created artificial "lightning, by passing [a] wire, in the dark, over a china plate that has gilt flowers."

One of Franklin's partners, Ebenezer Kinnersley, grew so adept at these kinds of exhibitions that, in need of work, and with Franklin's blessing, he packed up some of their electrical equipment and set off in 1749 on a lecture tour of the colonies. Broadsides for Kinnersley's appearances promised "A bright Flash of real Lightning darting from a Cloud in a painted Thunder-Storm"; "A Flash of Lightning made to strike a small House, and dart towards a little Lady sitting on a Chair"; and "A curious

Machine put in Motion by Lightning, and playing various Tunes on eight Musical Bells." It was advertised that the lecturer would present his audience with "Electrified Money, which scarce any Body will take when offer'd to them." Electrical experiments proved as popular as entertainments in America as they had in Europe. Samuel Domien, another acquaintance to whom Franklin taught some electrical tricks, later wrote his benefactor from South Carolina to say "that he had lived eight hundred miles upon Electricity [and] it had been meat, drink, and cloathing to him."

Franklin enjoyed the novelty of electricity as much as anyone, but was unsure what, if any, useful purpose it had. He set out to explore its potential for slaughtering turkeys and other small animals (eating his "victims" himself to note the tenderness of their meat). One group of pigeons he shocked seemed at first unaffected by the electrical charge until he noticed them milling about aimlessly in his courtyard and realized they had been blinded. He also dabbled in medical electricity, testing the effect of repeated shocks on human beings—"lunatics," "palsies," and "the melancholy"—although the physical effects from his electrical treatments never seemed to last longer than a few hours. He ultimately came to believe that when patients did report positive results, it was usually due to their own wishful thinking.

Franklin had already contributed far more than he imagined to the nascent science of electricity, although he assumed that more learned men of science in Europe must have by now outpaced him in electrical knowledge. In August 1747, he grew suddenly concerned that he had been too hasty in posting his findings to Collinson. "On some further experiments . . . I have observed a phenomenon or two that I cannot at present account for on the principle laid down in those letters, and am therefore become a little diffident of my hypothesis and ashamed that I expressed myself in so positive a manner," he wrote. "In going on with these experiments how many pretty systems do we build which we soon find ourselves obliged to destroy!" Franklin closed by requesting that Collinson not circulate any of his letters on electricity, or if he did so, to keep his authorship a secret. "If there is no other use discovered of electricity, this however is

something considerable," he sighed, "that it may help to make a
vain man humble."

Perhaps, given this momentary failure of confidence about his scientific work, it was a blessing that an urgent demand arose for
Franklin's attention to a local political crisis. Philadelphia, capital
of a British colony, had long felt secure from the danger posed by
marauding ships of hostile countries because it was almost ninety
miles inland from the Atlantic on the Delaware River. The existence of this buffer zone, like Pennsylvania's policy of situating
friendly Indian tribes between Philadelphia and potentially more
hostile natives farther west, suited the colony's Quaker leadership, which for religious and philosophical reasons had traditionally given little thought to military preparedness. In July 1747,
however, French and Spanish privateers attacked a ship from An-
tigua on the Delaware only twenty miles below Philadelphia,
wounding its captain. In another incident, a landing party raided
two plantations near Bombay Hook, stealing slaves and supplies.
Fear spread that an attack on Philadelphia was imminent.

From his desk in the Pennsylvania Assembly, Franklin saw
that factional infighting was likely to keep the body from re-
sponding adequately to the crisis, so he stepped forward to organize the city's self-defense. He wrote and published a pamphlet
entitled "Plain Truth" that stressed why an assault on the city
would not only be bad in and of itself, but might incite Indians on
the frontier to then take advantage of Philadelphia's sudden vulnerability. He reminded readers that the groups represented in
the city's population, such as the English, Germans, and Scottish,
had fierce war-making traditions, and that Quaker pacifism should
not be allowed to inhibit the townspeople, especially the merchants, from anteing up to buy arms and outfit a militia. He managed somehow to do this without alienating the Quakers, who
seemed willing to go along if not challenged directly over their beliefs. He then called for the formation and support of a volunteer
association of defense, arranged for a series of lotteries to raise

money, and, when not enough guns could be purchased on such short notice, negotiated with neighboring New York to borrow fourteen cannon, Franklin himself going to see the weapons brought safely overland through New Jersey. He even designed the heraldic emblems for the standards of the volunteer companies and began selling muskets and manuals of arms from his printing shop. Drawing on his New England background, Franklin also suggested that a day of fasting be proclaimed to "implore the Blessing of Heaven on our Undertaking." It was the first public fast ever encouraged in Pennsylvania, and would seem at odds with Franklin's Deistic views. But throughout his life he acted on the faith that, when the chips were down, God *would* intervene in human affairs. The proclamation of the fast, which he personally drafted and arranged to be translated into German, helped give local clergy a way to motivate their congregations to the cause of the town's defense.

The dreaded assault on Philadelphia never came, but his fellow citizens took note of Franklin's cool head in the emergency and his ability to transform fear into aroused public sentiment and action. Franklin was "the principal Mover and very Soul of the Whole," James Logan wrote, and all that was accomplished he did "without much appearing in any part of it himself." He had even published "Plain Truth" anonymously, signing it simply "A Tradesman of Philadelphia."

Franklin's exertions, and their good impact, presaged the far greater political demands that would soon take him away from Philadelphia and into the wider world. The theme of "Plain Truth" would resurface in 1754 as part of a plan Franklin put forward to unite all the American colonies, and the modest yet assured title would resound in a better-known revolutionary tract whose publication Franklin encouraged, Thomas Paine's *Common Sense* (1776).

Returning to his electrical investigations, Franklin and his colleagues focused on the potential of the Leyden jar, whose power continued to amaze natural philosophers. Franklin was the first to discover that the jar's stored charge was not in the water, as others had believed, but in the glass. The glass was a dielectric, meaning it stored and allowed the passage of electricity but did not conduct it. Having found that it was the glass of the Leyden jars that stored the charge, not the water, Franklin created a different sort of storage facility—a set of eleven panes of glass arranged vertically in close proximity, which he named an electrical "battery." The naming was a playful allusion to a rampart of guns, since the panes of glass stored and then "fired" a charge. About this time Franklin also coined the word "electrician," which was what he called himself.

When Franklin reported again to Collinson in spring 1749, it was to say that he and his cohorts were "chagrined a little that we have been hitherto able to produce nothing in this way of use to mankind." He added, in further self-deprecation on the theme of the futility of electrical investigations, that, with the humid summer approaching, electrical experiments would not be "so agreeable," and so he and the others had decided:

> To put an end to them for this season, somewhat humorously, in a party of pleasure on the banks of Schuylkill. Spirits, at the same time, are to be fired by a spark sent from side to side through the river, without any other conductor than the water: an experiment which we some time since performed, to the amazement of many. A turkey is to be killed for our dinner by the electrical shock, and roasted by the electrical jack, before a fire kindled by the electrified bottle; when the healths of all the famous electricians in England, Holland, France, and Germany are to be drank in electrified bumpers, under the discharge of guns from the electrical battery.

Although the arrival of warm weather did reduce static electricity in the atmosphere and inhibit tabletop electrical experimentation, Franklin, perhaps after witnessing one or two spring thundershowers, returned to his thoughts about bodkin points and flashes in the dark, and began to grow more curious about the possible similarity between electricity and lightning. Actually, he was fairly certain that lightning was electrical, but wondered if it was the same *kind* of electricity he knew firsthand. He wrote on April 29,

1749, to his Virginia acquaintance Dr. John Mitchell, suggesting a link between the electric discharge and lightning based on his investigation of the drawing off of an electric charge with points. Franklin's letter, which he titled "Observations and Suppositions, Towards Forming a New Hypothesis, for Explaining the Several Phenomena of Thunder-Gusts," is something of an electrical cosmology, offering a comprehensive account of how electricity is formed in the oceans by the confluence of water and salt, then rises into the atmosphere, where it electrifies the clouds, which then return the water to nurture the earth and its growing things. Although Franklin himself would soon acknowledge that he had the mechanism wrong, the theory represented an interesting conceptual leap, as Franklin tried to imagine how the entire atmosphere became charged with electricity.

Over the summer, he watched the thunderstorms that stirred and rumbled over the Delaware and read dispatches from other colonies of the damage to houses, churches, trees, and human beings. On November 7, 1749, he attempted again to organize his thoughts on the subject, jotting in notes that he later sent to Charleston physician John Lining that "Electrical fluid agrees with lightning in these particulars.

1. Giving light.
2. Colour of the light.
3. Crooked direction.
4. Swift motion.
5. Being conducted by metals.
6. Crack or noise on exploding.
7. Subsisting in water or ice.
8. Rending bodies it passes through.
9. Destroying animals.
10. Melting metals.
11. Firing inflammable substances.
12. Sulphureous smell.

In his workshop he called the communication between conductors of electricity "the spark and snap." Now he suggested that in "the great operations of nature" the "spark" would be lightning and the "snap" would be thunder. "If two gun barrels electrified will strike at two inches' distance, and make a loud snap," Franklin mused, "to what a great distance may ten thousand acres of electrified cloud strike and give its fire, and how loud must be that crack?" He had observed that a sharp needle would draw an electric charge from an electrified tube at a distance of twelve inches, while a blunt object would do so at three inches. Franklin noted that it was "constantly observable in these experiments" that the more electricity on the tube, "the farther it strikes or discharges its fire, and the point likewise will draw it off at a still greater distance." In one prescient exercise, he found that a charged pair of scales (simulating an electrically charged cloud), set at a height of two feet above a stationary iron object on the floor, would lose their charge if a sharp needle was held nearby, the needle drawing off the electrical attraction and preventing the charged side of the scale from descending. What Franklin was doing was inserting a doll-sized lightning rod into a miniaturized landscape. "The electric fluid is attracted by points," he concluded. "We do not know whether this property is in lightning. But since they agree in all particulars wherein we can already compare them, is it not probable they agree likewise in this? *Let the experiment be made.*"

Franklin, growing excited as the concept of the lightning rod began to take shape in his thoughts, assured Collinson, "There is something . . . in the experiments of points, sending off or drawing on the electrical fire, which has not been fully explained. . . . For the doctrine of points is very curious, and the effects of them truly wonderful; and from what I have observed on experiments, I am of the opinion that houses, ships, and even towers and churches may be effectually secured from the strokes of lightning by their means; for if, instead of the round balls of wood or metal which are commonly placed on the tops of weathercocks, vanes, or spindles of churches, spires, or masts, there should be a rod of iron eight or ten feet in length, sharpened gradually to a point like a needle, and gilt to prevent rusting, or divided into a number of points, which would be better, the electrical fire would, I think, be drawn out of a cloud silently, before it could come near enough to strike; and a light would be seen at the point, like the sailors' corpuzante."

The "corpuzante" to which Franklin referred, or *corpo del santo*, literally "the body of the saint," was commonly known as Saint Elmo's fire, a ghostly electrical phenomenon often occurring at sea, the name a familiarization of Saint Erasmus, the patron saint of the Mediterranean. Technically, the effect, which occurs when electrification of the atmosphere builds up around an insulated piece of metal, is called "electric induction," but it is more often referred to as a glow-discharge or brush discharge. This electrification causes the oppositely charged electricity within a metal rod to come to the rod's top; if the tension between the two is strong, the rod's electricity will discharge upward to neutralize the opposing force, creating a fine discharge of electrically excited gas molecules known as ions.

Mariners called the appearance of a single brush discharge aboard ship "Helena" and considered it a bad omen; the appearance of two, which they named "Castor and Pollux," augured good weather and a safe voyage. The mysterious appearance of the glow, especially in the pitch black of a moonless night in mid-ocean, stirred a humbled reverence in Christian sailors, who considered the dancing ions to be a holy apparition of Saint Elmo. Tears, fervent prayer, and other strong manifestations of faith were the common response to such a visitation. Pliny, centuries earlier, had marked how the discharge seemed to come to life. "I have seen myself in the camp, from the soldiers' sentinels in the nightwatch, the semblance of lightning to stick fast upon the spears and spikes set before the rampart. They settle also upon the cross sail-yards, and other parts of the ship, as men do sail in the sea: making a kind of vocal sound, leaping to and fro, and shifting their places as birds do which fly from bough to bough." The exact meaning of this mystery, Pliny conceded, was "hidden with the majesty of Nature, and reserved within her cabinet."

Despite the intriguing lore of Saint Elmo's fire, Franklin knew that the most dramatic evidence of the electrification of the atmosphere was not the brush discharge, but thunder and lightning. In an account that was sent to Franklin of a ship at sea struck by lightning, the ship's captain, John Waddell, reported that large "Comazants" appeared on the mastheads that "burnt like very large torches." A moment later the ship was struck by lightning, damaging the masts. Franklin surmised that the "spintles at the topmast-heads" were acting like a pointed bodkin and were drawing electricity from the clouds, and he noted that had there been "a good wire communication" between the masts and the sea, the wire would have channeled the lightning bolt harmlessly into the water, instead of allowing it to do great damage by encountering only ropes and wooden masts, which were nonconductors of electricity. Franklin had gained an increasingly comprehensive understanding of how lightning functioned as an electrical phenomenon, although, with his habitual caution, he told Collinson he still did not feel completely sure of his conclusions regarding points. But, he said, "as I have at present nothing better to offer in their stead, I do not cross them out: for even a bad solution read, and its faults discovered, has often given rise to a good one in the mind of an ingenious reader."

Despite the occasional stumble or hesitation, Franklin's simple love of experimentation generally saved him from ever becoming seriously incapacitated by doubt. Even though he could not yet explain lightning, he knew where his thoughts about bodkins and corpuzantes were leading—a device that might safeguard lives and property from lightning. "May not the knowledge of this power of points be of use to mankind," he asked, "in preserving houses, churches, ships, &c. from the stroke of lightning, by directing us to fix on the highest parts of those edifices, upright rods of iron made sharp as a needle, and gilt to prevent rusting, and from the foot of those rods a wire down the outside of the building into the ground, or down round one of the shrouds of a ship, and down her side till it reaches the water? Would not these pointed rods probably draw the electrical fire silently out of a cloud before it came nigh enough to strike, and thereby secure us from that most sudden and terrible mischief?"

When, in fall 1749, Franklin had written the words "*Let the experiment be made*," meaning a proof that would show if atmospheric electricity was the same as the sparks in his workshop, he had not yet determined what that experiment might be. By July 1750, however, he had hit upon what came to be called the "sentry box experiment." The idea was to place on top of a houselike structure that stood at a considerable height, for example on a hilltop, an elongated, pointed iron rod that would draw down an electric charge during a storm. The rod, positioned on top of the house, would poke down through the roof. The experimenter, or "sentry," seated inside, would hold a wire that touched the bottom of the rod, and would thus know when an electric charge came into the room. Seated on an insulated surface, and using a handle made of wax to hold the wire, the experimenter would be able to observe the spark without receiving a shock.

Given what Franklin knew about the deadly "mischief" lightning caused when it struck objects on the earth, his "sentry box experiment" appears somewhat casual about the use of a human investigator. However, he believed that the danger would be minimal if the experimenter remained alert; he attributed his own painful encounters with electricity, such as the time he had nearly electrocuted himself while attempting to slaughter a Christmas turkey, to his own carelessness.

Collinson, despite Franklin's modesty, had by now begun sharing Franklin's letters on electricity with others in England. Franklin, curiously, given his forgiving nature and his initial reluctance about the letters being seen, always nursed a tiny grudge against the Fellows of the Royal Society for not taking his initial thoughts about electricity more seriously. In 1788, two years before his death, he was still complaining (in the *Autobiography*) that his writings on electricity and the sameness of sparks and lightning had been "laught at by the Connoisseurs." Why Franklin believed this is unclear, for his letters were much admired and were even quoted by William Watson, England's foremost electrical scientist. Read before the Society, they were applauded for their "Clear Intelligent Stile" and the "Novelty of the Subjects." The Society had, it so happened, recently made known its wish that scientific articles not be weighed down by "a glorious Pomp of Words," and encouraged letter writers and contributors to communicate their ideas in a "close, naked, natural way of Speaking." Here Franklin's experience as a journalist and almanacker who wrote for the average reader served him well, enabling him to render scientific information in clear, straightforward prose.

One member of the Society who profoundly appreciated Franklin's achievement was John Fothergill, the Quaker physician who was himself famous for having written a highly readable book on the curing of sore throats. He recommended to Collinson that Franklin's letters be published, and at Fothergill's urging, Collinson gave the letters to Edward Cave, publisher of *Gentleman's Magazine*, to be assembled for publication.

Collinson may have acted promptly on Fothergill's suggestion because of his unhappy experience with another Philadelphia inventor, Thomas Godfrey. In 1730, Godfrey, an uneducated glazier and self-taught mathematician who was a protégé of James Logan, had devised a mariner's quadrant superior to those then in use; it employed two mirrors instead of one, and was effective even on the bobbing deck of a ship at sea. While Godfrey was still testing his instrument, however, a Londoner named John Hadley came before the Royal Society with a new quadrant whose design was similar to Godfrey's. Logan, although he had no proof, suspected that a prototype of Godfrey's quadrant that had been used by a ship's captain sailing for the British West Indies had either wound up in Hadley's hands or had been described to him. Despite the ardent efforts of Logan and Collinson, whose help Logan enlisted, the Society ultimately recognized Hadley as the quadrant's creator, giving him an award and voting Godfrey a consolation prize of two hundred pounds. (Informed that Godfrey would likely drink away the money, the Society eventually sent an ornate clock instead.)

Franklin, who at first had felt uneasy when he'd learned that Collinson was sharing his electrical observations with "experts" in London, now slowly accepted the possibility that they merited publication. "Wee thought it a great Pitty that the Publick should be deprived the benefit of so many Curious Experiments," Collinson assured him. With the issuance of his first major scientific text looming, in July 1750 Franklin composed a paper he called "Opinions and Conjectures, Concerning the Properties and

Effects of the Electrical Matter, Arising from the Experiments and Observations, Made at Philadelphia, 1749," a general summary of all his findings that would serve as the final chapter in the book Edward Cave would publish, *Experiments and Observations on Electricity, Made at Philadelphia in America, by Mr. Benjamin Franklin.*

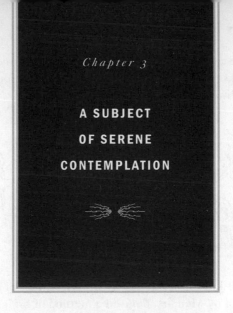

Chapter 3

A SUBJECT
OF SERENE
CONTEMPLATION

TODAY MANY SCIENTISTS SUSPECT LIGHTNING OF HAVING SPARKED the chemical evolution that began life on earth. So perhaps it is not surprising that thunder and lightning figure prominently in mankind's oldest recollections. It was once thought that lightning strikes were the result of visits to earth of great "thunderbirds" and that the gashes lightning left on trees were the scratches from their tremendous claws. In ancient Egypt, a god named Typhon was believed to throw down thunderbolts. In Greek myth it was Zeus sending flashes to earth whenever his favorite mortals got into trouble or someone stepped out of line. People in rural India spoke reverently of Indra, who carried a thousand-pointed thunderbolt on his cart while a crew of lackeys rode behind in a second vehicle, firing off individual flashes. Norsemen believed in the redheaded Thor, whose chariot wheels made thunder as he traversed the sky, forging lightning bolts with his magic hammer. For centuries, Scandinavian farmers whose plows turned up oddly shaped rocks in their fields assumed they had found relics of Thor's handiwork.

The best-known lightning myth may be that of Prometheus, who, assisted by Mercury, devised a method of bringing lightning down from the heavens at the end of a rod, and in this way stole fire from the sun, a crime for which he suffered an eternal punishment. According to the Latin poet Ovid, who lived at the time of Christ, monarchs sometimes claimed the power of drawing down lightning in the hope of impressing their subjects. In Roman myth, King Romulus showed his soldiers how to intimidate the enemy by slapping their swords against their armor to create "thunder"; Salmoneus achieved a similar deception by driving his chariot back and forth over a metal bridge. Both paid dearly for their trickery, as the gods, displeased at being so badly impersonated by mortals, smote each dead with a thunderbolt.

Watching the play of lightning in the Roman sky to divine the warnings or blessings of the gods was the duty of the College of Augurs, a body of distinguished older men who advised when important events like elections and coronations should be held. A lightning flash moving from left to right was a good omen; one going in the opposite direction augured poorly and might prompt a change of plans. (Of course, such a system was susceptible to abuse; an augur's promise "I will watch the sky" eventually became a euphemism for political deceit.)

The earliest "scientific" account of thunder and lightning was offered by Pliny, who thought lightning flashes were huge sparks emitted by the stars. "Like as out of a burning piece of wood a coal of fire flieth forth with a crack," he wrote, "so from a star is spit out as it were and voided forth this celestial fire, carrying with it presages of future things." Pliny attempted one of the first catalogs of lightning's habits, assuring his readers that a thunderbolt would not hurt anything more than five feet below ground, or disturb seals or eagles. Many such "verities" were spread well into the Middle Ages: Lightning poisons wine; animals killed by lightning must not be eaten; a tree struck by lightning should not be chopped up and used for firewood; a person struck and killed by lightning must be buried immediately. Lightning, it was said, at times came to earth in the form of a fiery ball capable of chasing people through their houses and then vanishing up the chimney. But it would not strike fig trees, the hyacinth, the laurel, grapevines, or onions; nor would it harm the eagle, hippopotamus, hyena, or crocodile.

Franklin himself entertained the fabled notion of what were called "lightning pictures." People or animals who had close brushes with lightning or were struck by it, went the theory, could be found afterward to have a mirrored imprint on some part of their flesh of the immediate surroundings at the moment of the flash. Franklin passed on to the French electrical experimenter Jean Baptiste Le Roy the story of a man who, standing in a doorway in a storm, saw lightning strike and was later found with a "counter-proof" of a nearby tree on his breast. It had appeared originally in Franklin's *Pennsylvania Gazette* on August 12, 1736:

> We hear from Virginia that not long since a Flash of Lightning fell on a House there, and struck dead a Man who was standing at the Door. Upon examining the Body they found no Mark of Violence, but on his Breast an exact and perfect (tho' small) Representation of a Pine Tree which grew before the Door . . . 'tis added that great Numbers of People came out of Curiosity, to view the Body before it was interr'd.

Such stories often involved crosses, other religious symbols, or even prayers becoming imprinted on clothes or flesh. In one seventeenth-century French account, lightning wrote in Latin on human flesh; in another it left behind "the sacred formula of the consecration of the mass." Occasionally it left images or messages that were invisible during the day but glowed radiantly at night. (Lightning itself was sometimes called "the light that turns the night to day," and certainly its ability to instantly whiten the darkest night as if it were broad daylight has always been one of lightning's most chilling and beautiful traits.) One of the best-known tales of a "lightning picture" occurred in the early nineteenth century near Bath, England, where lightning killed six sheep at pasture. It was said that when the animals' skins were displayed, viewers could identify the exact meadow where they died from the vivid imprint of the surrounding landscape.

As the lore of lightning pictures suggests, pagan notions about thunder and lightning were easily transmuted into early Christian beliefs. "It was, indeed, no great step," notes historian Andrew White, "for those whose simple faith accepted rain or sunshine as an answer to their prayers, to suspect that the untimely storms and droughts, which baffled their most earnest petitions, were the work of the arch-enemy, 'the Prince of the Power of the Air.' " Until the eighteenth century, when it ran up against Franklin's discoveries, Christianity freely propounded the notion that violent weather resulted from either providence or "diabolical agency," a belief sturdy enough to be embraced by numerous Catholic intellectuals, such as Saint Jerome, Saint Clement, Saint Augustine, and later, Martin Luther and Saint Thomas Aquinas, and represented in the "devil clouds" that show up in the margins of medieval books and in the verses of Dante. Luther credited a lightning bolt with setting him on his life's path. An indifferent young law student at the University of Erfurt in Thuringia, he was walking in the countryside on the afternoon of July 2, 1505, when he was caught in a sudden, violent thunderstorm. Confronting death and God's judgment, he fell to his knees and cried, "Help me, Saint Anne! I will become a monk." Two weeks later he entered the monastery of the Eremites of Saint Augustine.

Lightning, certainly the most fearsome arrow "diabolical agency" had in its quiver, could best be dispelled, it was said, by the ringing of church bells, preferably those engraved with Latin ad-

monitions and "baptized" with water from the River Jordan. The
church's special concern with lightning is perhaps easily understood. The tallest structure in eighteenth-century European or
New England towns was, invariably, a cathedral tower or church
steeple, and these structures, designed to scrape the clouds and
thus instill awe in the approaching worshipper, made ideal lightning targets. And because lightning travels through air, a poor conductor of electricity, it will change course the instant it detects a
superior route of conductivity between the ground and the sky,
such as a steeple. As an authoritative church history of Britain re-
counts, "there was scarce a great abbey in England which was not
burnt down with lightning from heaven," and several, including
the Monastery of Canterbury, the Abbey of Croyland, and the

Abbey of Peterborough, succumbed more than once and had to be rebuilt at great expense.

The "efficacious dispelling of aerial tempests" being of prime concern, a literature of special prayers and exorcisms was developed, including language for the "baptizing" of bells that would scare thunder away. "Whensoever this bell shall sound," promised one prayer of consecration, "it shall drive away the malign influences of the assailing spirits, the horror of their apparitions, the rush of whirlwinds, the stroke of lightning, the harm of thunder, the disasters of storms, and all the spirits of the tempest." Pope Gregory XIII, daring to address the bad weather devils directly, exclaimed in a widely quoted exorcism, "I, a priest of Christ . . . do command ye, most foul spirits, who do stir up these clouds . . . that ye depart from them, and disperse yourselves into wild and untilled places, that ye may be no longer able to harm men or animals or fruits or herbs, or whatsoever is designed for human use." A much-sought talisman against lightning was the Agnus Dei, the wax image representing the "Lamb of God." The fetish's power was believed to be greatly enhanced if consecrated by a pope in the first or seventh years of his pontificate.

Too respectful of religious faith to mock such beliefs openly, Franklin nonetheless found them amusing. Unfortunately, they could be deadly, particularly to the ringers of church bells. Church steeples, usually made of bricks or wood, were easily shattered by lightning, and the bell-ringers, racing up the stairs to combat the storm, were often themselves electrocuted, as the bell and its metal clapper attracted the fatal charge. In Germany, a 1784 publication entitled *A Proof That the Ringing of Bells During Thunderstorms May Be More Dangerous Than Useful* explained that, since 1750, a total of 386 church towers had been struck by lightning and 121 bell-ringers killed. On June 17, 1755, a lightning strike on one church had killed seven people instantaneously—three ringers in the church tower and four children hiding from the storm in the church below. The book's appearance led the Parlement of Paris, in 1786, to issue an edict prohibiting bell-ringing during storms, although the custom was so ingrained that as late as 1824 four new bells baptized against lightning were raised into place at the

Cathedral of Versailles, with government officials on hand to honor the event.

The idea of lightning as divine resentment—the well-known legal concept of "an act of God" probably originated with cases of death and destruction caused by thunderbolts—persisted well into the modern era, particularly in rural England and parts of southern Europe. But there have been more reasonable voices on the subject as far back as the Roman philosopher Lucretius, who in the first century B.C. asked why, if Jupiter was directing lightning bolts, so many landed in remote places deep in the forest, smashing obscure trees to bits. Martin Luther accepted the doctrine that devils caused bad weather but was skeptical enough to inquire why spirits powerful enough to hurl lightning would be frightened of church bells. Francis Bacon, who thought thunder was the sound of clouds falling on top of one another, doubted that bells would frighten the Devil but suggested that ringing them created some concussive energy that deterred lightning bursts, a view shared by René Descartes. This more naturalistic explanation of the bells' function was reflected in medieval bell-engravings such as "*Fulgura Frango*" ("I break up the lightning flashes") or "*Ego sum qui disipo tonitrua*" ("It is I who dissipate the thunders"). Some, like Father Augustine de Angelis, rector of the Clementine College at Rome, sought to better the odds by applying both notions simultaneously. "The surest remedy is the ringing of bells," he wrote in the late seventeenth century, "because the sound . . . by agitation disperses the hot exhalations and dispels the thunder. But the moral effect is the more certain, because by the sound the faithful are stirred to pour forth their prayers, by which they win from God the turning away of the thunderbolt."

Franklin came of age at a time when Father de Angelis's two-pronged strategy would have found numerous adherents, as naturalistic and providential explanations often intertwined. Newton himself, after all, believed the universe's functions to be of divine design. Lightning, wrote New England theologian Jonathan Edwards, could even be the subject of wonder, not fear, if it was understood as a purposeful element of the universe's elegant machinery. "Before, I used to be uncommonly terrified with thunder," he said, "but now, on the contrary, it rejoiced me. I felt God, so to speak, at the first appearance of a thunderstorm; and used to take the opportunity, at such times, to fix myself in order to view the clouds, and see the lightnings play."

Samuel Wall, a minister who in 1708 performed "experiments on the luminous qualities of amber, diamonds, and gum lac" before the Royal Society, reported that the "cracklings and light" from a piece of rubbed amber "seems, in some degree, to represent Thunder and Lightning." Newton, in 1716, wrote of being "much amused by the singular [phenomena] resulting from bringing a needle into contact with a piece of amber or resin fricated on silke clothe. The flame putteth me in mind of sheet lightning on a small—how very small—scale." In 1734, Stephen Gray noted that "the electric fire" was "of the same nature with that of thunder and lightning," and the abbé Nollet wrote in 1748, "If someone would undertake to prove that lightning is, in the hands of Nature, what electricity is in our hands . . . I admit that this idea, if well sustained, would please me greatly."

Franklin, with his description of his sentry box experiment, was the first person to suggest a means of testing this hypothesis. His aim was to solve a scientific enigma. But he knew full well that if he was right, and his proof disclosed that electricity was the actual source of lightning and thunder, he would be pulling back the veil on a subject surrounded for centuries by theology, fable, and superstition.

Franklin was relatively free to follow his intuition about lightning, even if it meant casting doubt on the popular faith in its divine origin. This was a result of his belief in the virtue of scientific curiosity and his "Aversion to arbitrary Power," certainly, but also because he lived in the New World, in Philadelphia, the most cosmopolitan community in North America. Unlike Europe, where natural philosophers still labored under the fear that their theories might offend church or state rulers and bring retribution in the form of censorship or even exile, in Franklin's colonial America, ecclesiastical authorities, especially outside New England, did not exert tremendous power. American science itself, with its emphasis on practical subjects such as botany, corn hybridization, and clock-making, was still decidedly homespun, and the New World's astronomers and academic men of science were far too few and dispersed to be perceived even remotely as a threat.

Of course, Franklin's personal conception of man's relationship with God must also be considered. A Deist who believed that God essentially left man alone, intruding only in the most dire circumstances, Franklin was "much disposed to like the World as I find it. I see so much Wisdom in what I understand of its Creation and Government, that I suspect equal Wisdom may be in what I do not understand: And thence have perhaps as much Trust in God as the most pious Christian." None of his neighbors, however, would ever have thought him pious. Franklin respected how religion served as a social adhesive and enhanced life for those bound to a faith, but he preferred generally to spend Sundays at home reading.

Franklin "had little more faith in orthodox doctrine than in witchcraft or astrology," according to biographer Alfred Aldridge, "yet he sympathized with the church as a social institution and supported it so loyally that many sectarians identified him with their causes." His Deistic beliefs were tempered by his concern that if, after all, God had really abandoned mankind, there would remain no standards of morality, only murkiness. "If men are so wicked as we now see them *with religion*," Franklin asked, "what would they be *if without it*?" The central tenet of Franklin's faith, then, if it can be characterized as such, was that the divinity must be essentially benign. He often quoted an epigram of Cato that he'd found in the writings of Joseph Addison:

> Here will I hold—If there is a Pow'r above us
> (And that there is, all Nature cries aloud,
> Thro' all her Works), He must delight in Virtue
> And that which he delights in must be Happy.

The concept of happiness, a state of being that relied not on wealth, property, or religious devotion but rather on an inner sense of fulfillment and self-determination, was new in the eighteenth century. It was codified in that ultimate document of the late Enlightenment, the Declaration of Independence, where the "pursuit of happiness" joined "life" and "liberty" as essential and "unalienable rights." Franklin, it was said, had "a talent for happiness." Certainly his philosophy of life departed thoroughly from hoary doctrines of a vengeful, punishing God. In a letter to a Connecticut-born friend, Jared Ingersoll, written while he toured Holland in summer 1761, Franklin said teasingly:

> I thought of your excessively strict observation of Sunday; and that a man could hardly travel on that day among you upon his lawful occasions without hazard of punishment; while, where I was, every one traveled, if he pleased, or diverted himself in any other way; and in the afternoon both high and low went to the play or the opera, where there was plenty of singing, fiddling, and dancing. I looked round for God's judgments, but saw no signs of them. The cities were well built and full of inhabitants, the markets filled with plenty, the people well favoured and well clothed, the fields well tilled, the cattle fat and strong, the fences, houses, and windows all in repair, and no Old Tenor anywhere in the country; which would almost make one suspect that the Deity is not so angry at that offence as a New England justice.

It would be a mistake, however, to conclude that Franklin did not dwell seriously on religious matters. Precisely the opposite was true. Others, settling early in life into a habit of reverence and church-going, surely became more devout; Franklin, standing outside organized religion like a respectful tourist, was endlessly intrigued by the power of prayer and by the vast diversity of sects and theologies. Thus released from any specific doctrine, he delighted in making speculations, some quite unorthodox, about the relationship between God and man. If the creator God had departed, Franklin wondered, where had he gone, and was he even still alive? Perhaps there were other gods, lesser or greater than the original, as various as the stars in the sky, who, like subcontractors, provided specific services to mankind; perhaps they were more approachable than the Creator. All that Franklin could honestly conclude was that God's nature being inaccessible to human reason or emotion, humans must symbolically represent it to themselves. This urge was based on the need to feel connected to the unknowable force that had given them existence and that they believed governed their lives and surroundings. To Franklin, any and all paths traveled in this difficult quest were equally viable; each was fumbling and uncertain, the sheer number of varying creeds convincing evidence that all methods of approaching the unknown must be "true." As if to prove the interchangeability of religious belief, one of his favorite hoaxes to play on friends was to pick up and pretend to read from the Bible a credible-sounding tale urging human tolerance, one that mimicked biblical language but that in fact he himself had made up.

The one attempt Franklin made as an adult at consistent church-going, in fall 1728, left him unsatisfied. Organized worship was mostly form with no substance, he decided, fine and finer readings of religious dogma that did not "inspire, promote or confirm Morality," but "serv'd principally to divide us and make us unfriendly to one another." Characteristically, he was moved by his disappointment to take matters into his own hands, writing out a creed of his own devising, his "Articles of Belief and Acts of Religion."

There was, he wrote, "one Supreme most perfect Being, Author and Father of the Gods themselves," a being "infinite and incomprehensible." All that could be assumed about so remote and unknowable a God was that he must have "in himself some of those Passions he has planted in us, and that, since he has given us Reason whereby we are capable of observing his Wisdom in the Creation, he is not above caring for us, being pleas'd with our Praise, and offended when we slight Him, or neglect his Glory. I conceive for many reasons that he is a *good* Being, and as I should be happy to have so wise, good and powerful a Being my Friend, let me consider in what Manner I shall make myself most acceptable to him."

Because Franklin thought God wanted humans to be happy,

and that happiness was impossible without virtue, he concluded that it was important to live a virtuous life. He faulted most religions for failing to emphasize this sufficiently. As *Poor Richard's* observed, "Serving God is doing good to man, but praying is thought an easier service, and therefore more generally chosen." When Franklin's parents mourned his apostasy, he assured them that "at the last Day, we shall not be examine'd [by] what we *thought*, but what we *did;* and our Recommendation will not be that we said *Lord, Lord,* but that we did GOOD to our Fellow Creatures"; similarly, in a 1753 letter to an acquaintance, he stressed that "making long Prayers, fill'd with Flatteries and Compliments" was likely "much less capable of pleasing the Deity" than "Works of Kindness, Charity, Mercy, and Publick Spirit."

Franklin's musings about faith, and his belief that man had the right to seek his own connection with the Creator, were no doubt inspired in part by the Great Awakening, the religious revival that electrified the colonies in the late 1730s and early 1740s. The religious and social uniformity that had characterized seventeenth-century New England was sent into retreat as evangelical preachers like Jonathan Edwards, the itinerant ministers James Davenport and Gilbert Tennent, and later the Anglican/Methodist George Whitefield stormed at audiences on Boston Common, throughout western Massachusetts and New Jersey, and as far south as Georgia, exhorting them to throw off their adherence to clerical elites and received truths and to discover God's grace and beauty in their own individual lives. The Awakening, a reaction to the rationalist trend of the times, reiterated the Calvinist message that one could be saved only by God's grace, yet emphasized the experience as a personal transformation. The extemporaneous thundering of the Awakening's "New Light" preachers, who sermonized without notes (as opposed to the rote, text-bound sermons of the Puritan "Old Light" clerics), was meant to shock listeners into a "New Birth."

The revivals were populist in tone, as listeners were encouraged to experience, in full view of others, a forceful emotional response to the ministerial exhortations. Congregants wept, fainted, or cried out, and some entered trances or claimed to have visions;

"multitudes were seriously, soberly and solemnly out of their wits," concluded New Haven's Ezra Stiles. Anyone who was "saved" was immediately empowered to preach God's lesson themselves, and numerous individuals—including, to the shock of church elders, several women—became fiery preachers, so imbued were they with what the disapproving Old Lights termed "enthusiasm," a word that implied an unacceptable degree of fervency. Some historians link the Awakening's insistence on a direct relationship with God with the weakening of social conventions in the colonies, or even cite it as an antecedent to the American Revolution. "Free choice," writes Alan Taylor, "had radical implications for a colonial society that demanded a social hierarchy in which husbands commanded wives, fathers dictated to sons, masters owned servants and slaves, and gentlemen claimed deference from common people."

George Whitefield, whom Franklin befriended in Philadelphia, was the movement's acknowledged luminary, "perhaps the most charismatic man to speak the English language during Franklin's lifetime." It was said he could make audiences gasp simply by pronouncing the word *Mesopotamia*. Diminutive and somewhat cross-eyed, Whitefield was so modest in appearance it was readily believed among his followers that only God could have given so plain a person such majestic powers of oratory and persuasion.

Whitefield first became known by touring England, offering his message of deliverance not in churches or cathedrals but among "the common people," in open spaces where the upturned faces of his listeners stretched a full quarter mile. In 1739 he arrived in America, expecting "pleasant hunting for sinners." "He was at first permitted to preach in some of our Churches," Franklin recalled, "but the Clergy taking a Dislike to him, soon refus'd him their Pulpits and he was oblig'd to preach in the Fields." Franklin became curious about the acoustics of Whitefield's outdoor sermons, which drew crowds estimated at six thousand or eight thousand. "[He] preach'd one Evening from the Top of the Court House Steps [and] I had the Curiosity to learn how far he could be heard, by retiring backwards down the Street towards the River." Franklin imagined "a Semi-Circle, of which my Distance should be the

Radius, and that it were fill'd with Auditors, to each of whom I al-
low'd two square feet." In this way, "I computed that he might well
be heard by more than Thirty-Thousand. This reconcil'd me to the
Newspaper Accounts of his having preach'd to 25,000 People in the
Fields, and to the antient Histories of Generals haranguing whole
Armies, of which I had sometimes doubted."

Franklin saw Whitefield as good copy, describing his activities
at length in the *Gazette* and publishing a number of his sermons.
And in the end he too succumbed to the minister's fabled charisma.

I happened . . . to attend one of his Sermons, in the Course
of which I perceived he intended to finish with a Collection,
and I silently resolved he should get nothing from me. I had
in my Pocket a Handful of Copper Money, three or four silver Dollars, and five pistoles in Gold. As he proceeded I
began to soften, and concluded to give the Coppers.

Another Stroke of his Oratory made me asham'd of that, and determin'd me to give the Silver. And he finished so admirably that I emptied my Pocket wholly into the Collector's Dish, Gold and all.

Always attentive to the related concepts of human character and reputation, Franklin was fascinated by Whitefield's methods. The minister was, well before the modern concept of it existed, a *celebrity,* inspiring others' intense adulation, using posters, ads, and other forms of promotion to attract crowds. Franklin, not dissimilarly, experimented with other new forms of communication and opinion-making, utilizing his newspaper, pamphlets, *Poor Richard's Almanac,* and the activities of his Junto and philosophical society to propagandize and spur his neighbors to action.

The two men, each exploring novel means of public engagement that fell outside traditional authority, thus no doubt saw a bit of themselves reflected in the other, and although far apart on many key issues of faith, remained steadfast friends their entire lives. Franklin approved of Whitefield's rejection of organizational doctrine and his dedication to good deeds, such as his generous efforts on behalf of orphans. Whitefield, meanwhile, could only have been impressed with Franklin as an example of a most remarkable secular man, one who gave unique thought to the meaning of devotion, and "whose dedication to public works and charity," Whitefield conceded, "had few Christian equals."

Isaac Newton had shown that gravity was a unifying, binding force between earth and the heavens. Franklin, in his book *Experiments and Observations on Electricity, Made at Philadelphia,* published in London in 1751, had suggested how yet another force of nature functioned. He had done so not by deduction or abstract calculation but by experiments, and in his clear commentary on the steps he had taken he had, in the best spirit of Newton's *Opticks,* helped "invent" an accessible written language of experimentation. Significantly, he called his book not "The Electrical Theories of Benjamin Franklin" or "The Laws of Electricity," but *Experiments and Observations*—his was an empiricist disquisition that wore proudly its own humble methodology.

For Franklin personally, of course, *Experiments and Observations on Electricity, Made at Philadelphia* represented a completed journey of ideas. Born in Boston only a decade after a local hysteria over spells and witches, he had been nourished on the words of English satirists and philosophers from far across the sea. Now, from remote North America, like the Gulf Stream current he would later study, he had sent back to Europe a quintessential Enlightenment tract that was welcomed as part of a crest of new rational inquiry. Montesquieu's *Spirit of the Laws* had arrived in 1748, with its sweeping recommendations for legal and social reforms and a humane form of government in which a system of checks and balances would diminish the tendency toward despotism. The next year came Denis Diderot's canny argument for open-mindedness, *Letter on the Blind* (which earned him a jail sentence). Jacques Turgot's *Discourse on the Successive Progress of the Human Spirit* was published in 1750, exhorting men to take heart because, while nature was perpetual and cyclical, humanity was infinitely capable of self-betterment; also published in 1750 was Jean-Jacques Rousseau's provocative *Discourse on the Arts and Sciences*, proposing that cultural refinement alienated men from their true natures and led societies to lose their moral grounding, and which brought Rousseau fame and a coveted prize from the Dijon Academy.

In 1751 appeared the first volume of the *Encyclopédie*, a veritable guide to the transformation the world was undergoing. The project, headed by Diderot and the mathematician Jean d'Alembert, would ultimately feature two hundred writers and three thousand illustrations, and run to twenty-eight volumes. It was intended to be not merely a repository of information, although it was certainly that, but "an engine of research and a stimulus to invention." It had been inspired in part by Ephraim Chambers's influential *Cyclopedia, or An Universal Dictionary of Arts and Sciences*, originally published in 1728, which pioneered the use of specialists to write articles and had been one of the first modern attempts at organizing all human knowledge. The entries from Chambers's book, tailored for a wide audience, were often excerpted in newspapers such as Franklin's *Pennsylvania Gazette*. The group of *philosophes* who contributed to Diderot's endeavor were known as *Encyclopédistes*, and included virtually every important man of letters in France. Turgot and Montesquieu wrote on history and politics, Buffon on natural history, Diderot on war, and Rousseau on music; other contributors included the Marquis de Condorcet and Voltaire, whose *Age of Louis XIV*, one of the first modern works of history, also hit the book stalls that year. The *Encyclopédie* itself described a *philosophe* as one who "regarded civil society as his 'divinity,'" and to whom Reason was what grace is to a Christian."

Franklin's book belonged fully to this moment. Reflecting the great public interest in electricity, the compact but sparkling little work, first appearing in London in April 1751, was immediately popular, and in translation soon became one of the most reprinted and talked about books of the age, blazing through five editions in English, three in French, one in German, and one in Italian. It was "the most important scientific contribution made by an American in the colonial period," according to historian Brooke Hindle. John Fothergill, in a preface he contributed, strongly praised the book's author as one who "conducts us by a train of facts and judicious reflections," and described electricity as "an invisible, subtile matter, disseminated through all nature," one that "if an unequal distribution is by any means brought about . . . becomes perhaps the most formidable and irresistible agent in the universe."

That Franklin's book of breakthrough science had been "made at Philadelphia" also represented something important: a statement of colonial pride and independence. Many Europeans still regarded America as a place of swamps, forests, and savages. Buffon's influential *Histoire Naturelle* had gone so far as to warn that conditions in the New World were so debased, immigrants might be assaulted by disease and laid low virtually upon arrival. John Singleton Copley, an American Loyalist painter who moved to London before the outbreak of the American Revolution and never returned, agreed that "in comparison with the [English] we Americans are not half-removed from a state of nature." Londoners, in fact, knew Philadelphia best for its proximity to New Jersey, a

EXPERIMENTS

AND

OBSERVATIONS

ON

ELECTRICITY,

MADE AT

Philadelphia in *America,*

BY

Mr. BENJAMIN FRANKLIN,

AND

Communicated in feveral Letters to Mr. P. COLLINSON,
of *London*, F. R. S.

L O N D O N:

Printed and fold by E. CAVE, at *St. John's Gate.* 1751.

(Price 2*s.* 6*d.*)

place where prostitutes, swindlers, and other miscreants were often shipped by the English courts. Thus, Franklin's sudden renown was enhanced by his own humble provenance, that so impressive a stride in natural philosophy as the "discovery" of electricity had been achieved by an *American*, a self-described tradesman and "leather-apron man" with only two years of formal schooling.

While English scientists had closely followed Franklin's discoveries, the greater interest came from the French, who seemed particularly taken with his depiction of electricity as a single force that sought a condition of balance and equilibrium. Buffon, who was France's de facto "minister of natural philosophy," arranged for his friend the physicist Thomas François D'Alibard to translate Franklin's work into French. He also encouraged D'Alibard to be the first to try the sentry box experiment described in Franklin's book, in which a man would stand on an insulated surface and draw sparks from an iron rod during a thunderstorm. Curiously, none of the English experimenters, despite a head start in knowing of Franklin's proposed proof, had yet attempted to investigate it.

D'Alibard, acting on Buffon's suggestion, ordered a small shack built and, above it, an iron rod forty feet high, in the garden of his home in Marly-la-Ville, a village about fifteen miles from Paris. Retreating to the capital on business, D'Alibard assigned a neighbor, a retired dragoon named Coiffier, to keep an eye on the experiment. On May 10, 1752, at twenty minutes past two in the afternoon, the wind suddenly picked up, the rumble of thunder was heard, and an electrical storm descended on Marly-la-Ville. Resolving gallantly to try and obtain a spark in D'Alibard's absence, Coiffier climbed onto the insulated stool and held a brass wire leading from a Leyden jar to the bottom of the rod: All at once a shower of sparks flew over his hand. The old soldier cried for someone to go fetch Raulet, the local priest. Raulet came running, his haste alarming the other villagers, who assumed he was hurrying to administer the last rites to Coiffier, apparently killed by D'Alibard's strange "lightning-house." But "the honest ecclesiastic arriving at the machine and feeling there was no danger," explained a later account, "took the wire into his own hand, and immediately drew several strong sparks, which were most evidently of an electrical nature, and completed the discovery for which the machine was created."

D'Alibard reported the success three days later to the Académie Royale des Sciences in Paris, stating: "In following the path that M. Franklin traced for us, I have obtained complete satisfaction." Franklin's theory that thunderclouds are electrified was reaffirmed on May 18 when Delor, a Parisian electrical exhibitor (who went by one name), also successfully staged the sentry box experiment.

Word quickly spread of the accomplishment, flinging the door open on a vast new field of philosophic inquiry. If thunder and lightning were, as the sentry box experiment had shown, natural phenomena, they could be perhaps more closely investigated and measured. Throughout the spring and early summer, the proof was repeated and verified by numerous European electricians, including the abbé Nollet, two researchers in Germany, and, in England, by William Watson and another Royal Society Fellow, John Canton. On June 7, Louis XV's physician successfully drew sparks during a tempest at St. Germain-en-Laye, having situated a rod in the garden of the Hôtel de Noailles. On June 23 and 26 in Brussels, sparks were drawn from a rod affixed to the top of a house. D'Alibard, in an adaptation of his first effort, joined several iron bars together to create a rod nearly sixty feet long and then suspended it with silk cords so that it rested atop a row of glass containers. At the approach of a storm, the bars became so thoroughly electrified that the sparks almost set an assistant's coat on fire.

Praise for Franklin was immediately forthcoming. In solving one of the earth's great mysteries, he had stripped nature's most fearsome phenomenon of its mystical provenance. "The effect on the public mind was awe-inspiring, and can justly be compared with that later to be produced by the explosion of the first atom bomb," according to the twentieth-century lightning expert Basil Schonland. Joseph Priestley, the English scientist famous as the discoverer of oxygen, who also investigated electricity and wrote a seminal book on the topic, thought Franklin's proof "the greatest discovery that has been made in the whole compass of philosophy since the time of Sir Isaac Newton," and noted that many Europeans upon first hearing of "Mr. Franklin's project for emptying clouds of their thunder . . . could scarce believe him to be any other than an imaginary Being." (Considering that Franklin was seen in his own lifetime as a leading representative of the Enlightenment, it's ironic that many could not help regarding him as a kind of sorcerer. During the Revolutionary War, which Franklin spent in Paris as America's representative to the court of France, some Britons voiced the fear he might use his uncanny skill to command thunder to send some sort of electrical apocalypse across the English Channel.)

Along with their initial sensation, Franklin's ideas made possible a more permanent shift in "the earth sciences." Electricity had been thought of until recently as a "toy science," with its electrical kisses, dangling boys, and electrocuted turkeys, but now, as Cohen explains, "every experimenter rubbing glass tubes in his laboratory knew that he was studying cosmic forces on a small scale." Franklin's conclusions demanded that electricity join gravity, light, heat, and meteorology in any account philosophers offered for the majestic workings of nature. Thunder and lightning, which Watson and other scientists had long seen as "almost inaccessible to their inquiries," had overnight become, in the estimation of nineteenth-century lightning rod advocate William Snow Harris, "a subject of serene contemplation."

Of course, as it took news from Europe anywhere from four to eight weeks to cross the Atlantic to Philadelphia, Franklin himself remained unaware of D'Alibard's success with the sentry box and the tremendous stir it had caused. He had in fact developed an alternate plan of his own to execute the experiment, using the new spire of Christ Church near Market Street in Philadelphia, but he had been delayed because construction of the steeple, to which he had contributed funds, had not proceeded on schedule. The setback was frustrating because June was "thunder-gust" weather in Philadelphia, offering ideal conditions for lightning studies. As he watched clouds scroll across the sky behind the unfinished church tower, however, a new, perhaps easier, version of the sentry box experiment formed in Franklin's mind: If he could not use the church spire to bring electricity out of the sky, why not employ that most ordinary of children's diversions, *a kite*, to reach up into it? The idea, of course, was counterintuitive; everyone knew kites were flown in fair weather, and thrived on light, steady breezes. No one would, or probably ever had, purposely sent one aloft in a thunderstorm.

Franklin's kite experiment of June 1752 has long been the subject of mild controversy. He was secretive about carrying it out and shy about relating it, and the only purported eyewitness, his twenty-one-year-old son, William, never wrote or commented about it. When he did describe the experiment, Franklin was a bit more circumspect than usual. Various theories have been considered for his reticence: his habitual modesty, the fear that he would look ridiculous flying a kite in the rain, the death that year of his mother, Abiah, distracting him from his work, or even the possibility that he never carried out the experiment at all but merely thought it through to conclusion in his mind and, pleased with the "results," reported it as complete. Perhaps, closely engaged as he was with the subject, the discovery—in his eyes—lacked the grandeur and importance others assigned it; after all, he had been very certain lightning was electrical: His question was simply whether it was the same kind of electricity he was already familiar with.

Lacking definitive evidence, one can't help giving Franklin the benefit of the doubt. As fond as he was of playful hoaxing, science was something he revered too highly to misrepresent; and any insinuation that, learning of D'Alibard's triumph in Europe, he then fabricated the kite story in order to assert priority, also rings false. The date of D'Alibard's breakthrough was May 10, and Franklin reported having flown his kite in June. If his intention was to claim priority, why would he not have backdated the kite experiment to a time prior to May 10? And if claiming distinction for his role in the experiment really mattered to him in the first place, why did he publish instructions for the sentry box in a book, or commit to paper the exhortation, "*Let the experiment be made*"?

As a once popular Philadelphia ode narrates:

A heavy cloud above the city lay,
And fitful gusts, a coming storm foretold;

The merry urchins ceased their out-door play,
And gathered in the dear domestic fold.
Then Franklin, bent on noble sport and bold,
To mead adjacent, hastened with a kite . . .

While the rest of Philadelphia headed for cover that blustery afternoon, Franklin and William went "out on the commons," or "not far from his own house, say about the corner of Race and Eighth streets, near a spot where there was an old cow shed," depending on which version of the story one prefers. Another local variation places Franklin's legendary experiment a bit north of town, near the present intersection of Ridge Road and Buttonwood Street. An argument for the more remote location is that the Franklins wanted to avoid being observed flying a kite in a downpour. A respected contributor to a late-nineteenth-century compendium of Philadelphia history reported that, as a child, he had been taken to this spot and had had pointed out to him the remnants of the old shed that the father and son had used that day.

Franklin carried with him a kite he had made of silk and cedar. To the top of the upright stick he had attached a sharp pointed wire that rose a foot or more above the wood. The twine leading down from the kite was attached to a silk ribbon, and on the silk ribbon dangled a key. It was important that Franklin and William stand indoors because the silk ribbon must remain dry, and hence the need for the shed. There was a strong breeze, and with William's assistance, the kite was soon aloft; Franklin retreated to the threshold of the shed and waited, playing the kite in the direction of the oncoming storm.

For a long time he detected nothing, even as several dark, brooding clouds passed directly overhead. Then, suddenly, Franklin noticed the individual strands of hemp on the twine move as one and abruptly stand erect. He carefully brought his free hand forward and extended a knuckle to the key and received a mild shock. A moment later the clouds opened and the rain came on, dampening both kite and twine. Because electricity is more readily conducted through water and objects that are wet, it surged downward to the shed in a rush, causing sparks to stream off the key to Franklin's hand, allowing him to collect "electric fire very copiously."

The event, epochal as it was, may not have seemed particularly momentous to the experimenters. Most likely they waited until the storm let up, gathered their gear, and went home, where they endeavored to explain their damp clothes to a skeptical Mrs. Franklin. Franklin himself would take several months to write of the adventure, and a full account—presumably given by Franklin to Joseph Priestley—would wait until Priestley published his history of electricity more than twenty years later. The spark flying from the key to Franklin's knuckles, however, would soon, and in

Franklin's testing proved his point

To check his hunch that lightning was a form of electricity, he made a silken kite and flew it in a thunderstorm. When a charge ran down the rain-soaked cord, Ben's theory became an established fact.

his lifetime, be endowed with mythic importance—a seminal occurrence in the history of science and a radiant symbol of the Enlightenment. "At the moment when he drew the electric spark from the cloud," wrote historian Andrew White, "the whole tremendous fabric of theological meteorology reared by the fathers, the popes, the medieval doctors, and the long line of great theologians, Catholic and Protestant, collapsed; the 'Prince of the Power of the Air' tumbled from his seat; the great doctrine which had so long afflicted the earth was prostrated forever."

Despite the evidence provided by his kite experiment and the news of the sentry box demonstrations in Europe, Franklin remained a bit unsure of what, if any, practical application his discoveries would have. Now that it had been proven that lightning and electricity were the same, however, he could proceed with the theory he had been nurturing ever since he observed the point of a bodkin attract an electrical charge—the possibility that a rod of iron with a sharp point, situated atop a tall structure, would draw off the electricity from a thundercloud as a storm approached, thus rendering it harmless. If thunder and lightning were, as Franklin understood, an electrical communication between the sky and the earth, a transfer that restored the atmosphere's electrical equilibrium, then a pointed lightning rod would, by facilitating a brush discharge, ease the electrical tension in thunderclouds and spare damage to property and animals. He also speculated that, if properly grounded, the rod would conduct lightning safely to earth.

That summer of 1752, Franklin dwelled on lightning's dangers, featuring numerous accounts of lightning strikes in the pages of the *Gazette*. "We hear from Susquehanna, that on Sunday . . . a Man was struck dead by the Lightning; that another Man was so stunned that it was some time before he recovr'd; and that a child, who sat betwixt the legs of one of them, rece'vd no damage."

In Philadelphia, "a Bull and two cows were killed by Lightning . . . and tho' a woman was milking one of the cows at the same instant, yet she received little or no hurt."

Franklin was intrigued by the odd unpredictability of lightning, how it might kill a cow instantaneously yet leave a milkmaid unscathed, why a boy sitting between two men struck by lightning felt no effect even as one of the men fell dead. He was also curious about how lightning made its destructive path through a building. He saw that lightning followed a course of least resistance, seeking to complete its connection between the electrified cloud and the ground as hurriedly as possible, finding those materials that were highly conductive, and avoiding or simply destroying those that were poor or nonconductors of electricity. "It was very remarkable in both Houses," he reported on August 5, 1752, of lightning that struck two residences on Philadelphia's Society Hill, "that the Lightning in its Passage from the Roof to the Ground, seem'd to go considerably out of a direct Course, for the sake of passing thro' Metal; such as Hinges, Sash Weights, Iron Rods, the Pendulum of a Clock, &c. and that where it had sufficient Metal to conduct it nothing was damag'd; but where it passed thro' Plaistering or Wood work, it rent and split them surprizingly."

Franklin recognized that when lightning struck the earth it sought conductors, those materials that would take it instantly into the ground. Not finding adequate conductors, he explained, "the fluid passes in the walls whether of wood, brick or stone, quitting the walls only when it can find better conductors near them, as metal rods, bolts, and hinges of windows or doors, gilding on wainscot, or frames of pictures; the silvering on the backs of looking-glasses; the wires for bells; and the bodies of animals, as containing watry fluids. And in passing thro' the house it follows the direction of these conductors, taking as many in its way as can assist it in its passage." The reason lightning damage often seemed haphazard, he saw, was that lightning "will go considerably out of a direct Course for the sake of the Assistance of good Conductors." He specified that lightning's fierce electrical charge "is actually moving, tho' silently and imperceptibly, before the Explosion in and among the Conductors." The explosion occurs "only when the Conductors cannot discharge it as fast as they receive it, by reason of their being incompleat, disunited, or not of the best Materials for Conducting."

In September, for experimental purposes, he placed a nine-foot rod atop the chimney of his own house, then led a wire from it through a glass tube in the roof and down through a center staircase. Outside his bedroom door the wire split, and at each end he affixed a small bell. Between the bells, which were six inches apart, he ran a silk thread with a tiny brass ball that played the bells when electrified clouds passed overhead, somewhat to the chagrin of his wife, Deborah, who did not share Franklin's curiosity about lightning to the extent of wanting to draw it down into her family's abode. One can imagine her concern at the incident that prompted him to record:

> I was one night waked by loud cracks on the staircase. Starting up and opening the door, I perceived that the brass ball, instead of vibrating as usual between the bells, was re-pelled and kept at a distance from both; while the fire passed, sometimes in very large, quick cracks from bell to bell, and sometimes in a continued, dense, white stream, seemingly as large as my finger, whereby the whole staircase was enlightened as with sunshine, so that one might see to pick up a pin.

This charming homemade experiment—musical and scientific at once—pleased Franklin a great deal. A 1762 portrait of him (by Mason Chamberlin) at his desk, noting the action of the bells as a lightning rod and an approaching "thunder-gust" are seen through a window, was one of Franklin's favorites, and he often sent a popular engraving of it, titled "B. Franklin, Philadelphia," to friends and diplomatic acquaintances.

In summer 1752 Franklin put grounded lightning rods on both the Pennsylvania State House and the Pennsylvania Academy, the first erected anywhere in the world. That October he printed in the *Gazette* a brief account of his kite experiment, and in a letter to Collinson described "making a Machine or Kite" that, when flown in a thunderstorm, would draw "Electric Fire from the Clouds to such a degree as to charge a Phial, kindle Spirits, and perform all the other Experiments which are usually done by rubbing a Glass

Il a ravi le feu des Cieux ;
Il fait fleurir les Arts en des Climats sauvages.
L'Amérique le place à la tête des Sages
La Grèce l'auroit mis au nombre de ses Dieux ;

Globe or Tube, and thereby the Sameness of the Electric matter with that of Lightning may be demonstrated." The following month he used the pages of *Poor Richard's* for 1753 to describe his lightning rod and suggest how it should be installed. The announcement, one of the more unassuming mentions of a major technological development in scientific history, was tucked between workaday notices about upcoming Quaker meetings in several colonies and court dates in the new year at Annapolis and appeared beneath the heading "How to Secure Houses, etc., from Lightning."

> It has pleased God in His goodness to mankind at length to discover to them the means of securing their habitations and other buildings from mischief by thunder and lightning. The method is this: Provide a small iron rod (it may be the rod-iron used by the nailers) but of such length that, one end being three or four feet in the moist ground, the other may be six or eight feet above the highest part of the building. To the upper end of the rod fasten about a foot of brass wire the size of a common knitting needle, sharpened to a fine point; the rod may be secured to the house by a few small staples. If the house or barn be long, there may be a rod and point at each end, and a middling wire along the ridge from one to the other. A house thus furnished will not be damaged by lightning, it being attracted by the points and passing through the metal without hurting anybody. Vessels, also, having a sharp-pointed rod fixed on the top of their masts, with a wire from the foot of the rod reaching down, round one of the shrouds, to the water, will not be hurt by lightning.

One of the first fully documented accounts that lightning rods actually worked came from Philadelphia after Franklin had gone to England as agent for Pennsylvania. It involved the home of a local merchant named William West, who notified Franklin's collaborator Ebenezer Kinnersley that, in a recent storm, lightning had struck his house and had been safely borne to earth by the rod. "I waited on him to enquire what Ground he might have for such Suspicion," Kinnersley reported in a letter to Franklin. West explained there had been a deafening crack and a flash of lightning directly overhead, but no visible damage to the house had been found. Insisting that they ascend to the roof to examine the rod, Kinnersley found its point thoroughly melted. Franklin had by now backed away somewhat from his initial theory that a lightning rod would draw the pent-up electricity from a cloud, emphasizing

instead the rod's ability to attract the lightning bolt itself and conduct its charge into the ground. The evidence that Kinnersley
found on Mr. West's rooftop was, in Kinnersley's words, "convincing Proof of the great Utility of this Method of preventing [light-
ning's] dreadful [effects] . . . May the Benefit thereof be diffused
over the whole Globe. May it extend to the latest Posterity of
Mankind; and make the Name of FRANKLIN like that of Newton,
immortal."

One reason that Franklin became an overnight scientific celebrity—in 1753 he won the Royal Society's Copley Medal and received honorary degrees from both Harvard and Yale—was
because his work was so easily understood. People of virtually any
background and level of education could grasp the basic lightning–
electric spark analogy and the function of the lightning rod. But
not everyone was willing to accept Franklin's ideas, especially his
rooftop invention. Many questions and criticisms arose; some were
legitimate science, some knee-jerk theology, and a few simply expressed established science's perennial suspicion of audacious
newcomers.

The most concerted opposition came from the abbé Nollet,
ironically one of the few prominent European men of science
whose personal trajectory somewhat paralleled Franklin's. Born to
a peasant family in the village of Pimprez, in 1700, Nollet's preco-
ciousness had led the village curé to arrange for him to attend a
provincial college in Clermont and then advance to theological
study in Paris. Drawn to science, he decided against joining the
priesthood, although he retained the title he had earned at semi-
nary and later wore ecclesiastical garb at court, perhaps to help im-
munize his scientific views. Through membership in a Société des
Arts devoted to promoting natural philosophy among artisans, he
became a protégé to two of France's leading scientific scholars—
the electrical experimenter Du Fay and the entomologist René
A. F. de Réaumur. Too poor to purchase scientific instruments, Nollet
became skilled at building his own, and by the late 1730s emerged

as one of France's leading instrument makers, selling his machines to many foremost *philosophes*, including Voltaire. Known for his own great talent as an experimenter—it was said he could perform demonstrations on no fewer than 350 instruments—Nollet cut a notable figure in fashionable Paris, a handsome boy-wizard from the provinces. His meteoric rise may have contributed ultimately to his undoing, for he began to fancy himself an invincible paladin of the Enlightenment, dispelling lesser mortals' "vulgar errors, extravagant fears and faith in the marvelous."

Like his counterpart across the English Channel, William Watson, Nollet was inspired in 1745 by Georg Mathias Bose's experiments with spirits and sparks, and by the advent of Musschenbroek's Leyden jar. As an electrical showman, Nollet was one of the first to comprehend the tremendous potential of the bigger, more reliable charges made possible by the jar. It was now feasible for an experimenter to provide himself with a more or less continual brush discharge, a phenomenon that, compared to isolated sparks, could not help but suggest new hypotheses about electricity's nature.

Retiring to his laboratory to examine these effects, Nollet emerged with an idea that overnight made him the most famous electrician in Europe: a comprehensive theory of electricity as a corpuscular flow of matter. He concluded that electricity departs an object in divergent conical jets, the object's "effluence." Nearby objects, and the air itself, return an "affluence." The effluent flow is divergent, the affluent more unified, so electrical imbalances always exist, but both consist of the same electrical force and differ only in the direction in which they flow.

Nollet's unraveling began when he became the victim in a feud between Réaumur and the powerful Buffon. The two men had academic differences, as Réaumur was a devoted Cartesian and Buffon a man who presumed nature held some innate *sensibilité*. One of the most imaginative scientific minds of the era, Buffon also had a streak of petty vindictiveness. In his *Histoire Naturelle* he had belittled Réaumur's scholarship. Réaumur returned fire via a friend, Joseph-Adrien Lelarge de Lignac, whose *Lettres à un Américain* asserted that Buffon's "way of reasoning is even more revolting than his hypotheses." Buffon also suspected Réaumur of conspiring to get his friend Diderot sent to prison.

When Franklin's book on electricity appeared, Buffon saw a chance to get in some digs at Réaumur by discomfiting his electrical protégé, Nollet, whom he disliked anyway and thought overly proud, by arranging for a translation of the theories of Franklin in a way that would potentially embarrass the abbé. D'Alibard, in the foreword to his translation of Franklin's book, offered a survey of electrical knowledge that included the views of many lesser-known European physicists, and omitted all mention of Nollet. This alone would suggest to the abbé that someone was playing a hoax, and because he had never heard of Benjamin Franklin of Philadelphia, he couldn't help but wonder at first if someone had made him up. When he did accept that Franklin was real, he was perturbed to find the American getting all the credit for a theory linking electricity and lightning, as he had earlier recorded the same similarity. In a further humiliation, D'Alibard, along with Buffon and Delor, had obtained an invitation to Versailles, where they entranced king and court with demonstrations of some of Franklin's experiments, a bold incursion onto the turf of France's "court electrician." Louis XV was so impressed he had, probably at Buffon's suggestion, dictated a letter to the Royal Society communicating "the King's Thanks and Compliments in an express manner to Mr. Franklin of Pennsylvania." D'Alibard's experiment at Marly-la-Ville came soon after, and as no one recalled that Nollet himself had once drawn the parallel between electricity and lightning, and all the talk was of the Marly demonstration, and of Franklin, Nollet saw his great fame vanish as if in a cloud of sorcerer's smoke. "The abbé Nollet," confided Buffon to a friend, "is dying of chagrin from it all."

Nollet responded with a series of letters mocking and correcting Franklin's skills as an experimenter, an area where the veteran instrument-maker knew himself to be on solid ground. Franklin, at first willing to defend his ideas, ultimately chose not to respond. "When my papers were first published, the abbé Nollet, then high in reputation, attacked them in a book of letters," Franklin recalled. "An answer was expected from me, but I made none to that book, nor to any other. They are now all neglected, and the truth seems to be established. [I] concluded to let my Papers shift for themselves; believing it was better to spend what time I could spare from public Business in making new Experiments, than in Disputing about those already made." Franklin was generally true to his word, rarely, if ever, allowing feelings of personal proprietorship to interfere with what he viewed as the overarching quest for scientific truth. Indeed, he was known to praise the experiments of others even when their conclusions threatened his own. In the case of Robert Symmer, a skilled English investigator who challenged some of the tenets of Franklin's single fluid theory, it was Franklin who made available the equipment Symmer required to carry out the work.

Franklin also readily conceded his inability to say what caused the electrification of the atmosphere. This was a challenge to which he frequently returned, and one that must have been particularly frustrating to someone with Franklin's passion for knowing nature's systems. Having noticed that ocean water sometimes turns luminescent in the dark of early evening, he briefly promulgated in the early 1750s the theory that "the Sea might possibly be the grand Source of Lightning." Electrified by the friction of salt particles with water, he wrote, water evaporates and in its transformation to vapor electrifies the clouds, and the clouds then discharge their electricity to the earth. However, on direct experimentation he found that "Sea Water in a Bottle, tho' at first it would by Agitation appear luminous, yet in a few hours it lost that Virtue; hence, and from this, that I could not by agitating a Solution of Sea Salt in Water produce any Light, I first began to doubt of my former Hypothesis." Always willing to correct his own scientific ideas, on further investigation he had found that the phosphorescence of seawater is caused not by friction with salt but by floating animalcules, an idea first postulated by Linnaeus. He then speculated that air might become charged by its friction against the earth, a theory that also did not prove satisfactory, before ultimately conceding, "Perhaps some future experiments on vaporized water may set this matter in a clearer light."

While Franklin chose not to defend his views to Nollet, he was appreciative of others who did. Giambatista Beccaria, a professor of experimental physics at the University of Turin, repeated many of Franklin's experiments, including flying a kite in a thunderstorm, and helped introduce lightning rods in Italy. As in Buffon's case, Beccaria appears to have endorsed Franklin's theory partly as a means of shaking things up in the Italian sciences. In 1753 he produced a slim but potent work, *Dell'Elettricismo Artificiale e Naturale*, which carefully laid out Franklin's theories and echoed his observations about the parallels between laboratory electricity and the workings of the atmosphere. Beccaria's book, which included a "letter" to Nollet that dealt summarily with the abbé's ideas, was immediately taken up by Franklin's French devotees, the *franklinistes*—D'Alibard, Delor, and Jean Baptiste Le Roy, among others—and translated into that language.

Nollet, though always quick to claim priority for recognizing the electrical nature of lightning, was not keen on the lightning rod. As a man of the church, he deemed it "as impious to ward off Heaven's lightnings as for a child to ward off the chastening rod of its father," and as a scientist he was inclined to reject the notion that the overwhelming power of a thunderstorm might be conquered by a thin rod of iron mounted on a rooftop. In his series of letters written to refute Franklin, Nollet noted that "All these iron points . . . are more likely to attract lightning than to save us from it and . . . the idea of dissipating the charge in a thundercloud [by points] is not scientific." As proof, Nollet alluded to a case that had deeply unsettled the world's community of electrical researchers, the electrocution of the physicist Georg Wilhelm Richmann, killed in St. Petersburg on August 6, 1753, when a version of the Franklin sentry box experiment he had constructed took a direct lightning strike.

Richmann had come to Russia originally from his homeland in present-day Estonia to serve as a tutor to the children of aristocrats. He became a member of the Imperial Academy of Science in 1735 and, a decade later, a professor of experimental philosophy. Believing it "my business to enquire into nature so far as I am able, and to neglect no occasion not only for observing, but also for measuring the phenomena of natural electricity," he had created what he called "an electrical gnomon" or "electrometer"—a means of measuring not only the presence of atmospheric electrification but also its strength. The gnomon consisted of a glass vessel with brass filings at the bottom and an iron rod that extended down from a set of iron bars on the roof of Richmann's house into the vessel. A thread tied to the upper part of the rod lay flat alongside it except when the rod became electrified. When that occurred, the degree of the angle at which the thread stood away from the rod provided Richmann with a measurement of the intensity of the atmospheric electricity.

On August 6, Richmann was attending a meeting of the Academy when, shortly before noon, distant thunder alerted him to an approaching storm. He hurried home in the company of I. A. Sokolov, an engraver who was working with him to provide illustrations for a book the professor was preparing about his discoveries. Several minutes later, people on the street noticed an unusually dark cloud that "seemed to float very low in the air," and in a moment heard "such a thunder-clap, as has hardly been remembered at Petersburg." Whether this bolt struck directly the iron bars atop Richmann's house is unclear, but it fell close enough to send a substantial electrical force down through the roof to his measuring device. At that moment, Richmann had the misfortune to be peering closely at the action of the mechanism, and "as he stood in that posture, a great white and bluish fire appeared between the rod of the electrometer and his head." His companion, Sokolov, saw a "globe of blue fire" as large as his fist "jump from the rod of the gnomon towards the head of the professor, which was . . . at about a foot distance from the rod." Sokolov himself was knocked down by the blast, jumped up, and ran out of the house in a panic but then returned immediately to find the apparatus in ruins and Richmann sitting in a crumpled posture on a nearby chest. The professor's wife, who had hurried into the room after the explosion to find her husband "past sensation," was in tears, vainly pleading with him to awaken.

The first man to die from a lightning strike while studying lightning, Richmann himself became the object of immediate scrutiny. "They opened a vein of the breathless body twice, but no blood followed," Priestley wrote. "Upon turning the corpse with the face downwards, during the rubbing, an inconsiderable quantity of blood ran out of the mouth. There appeared a red spot on the forehead, from which spirited some drops of blood through the pores, without wounding the skin. The shoe belonging to the left foot was burst open, and, uncovering the foot at that place, they found a blue mark; from which it was concluded, that the electrical force of the thunder, having entered the head, made its way out again at the foot." Word having quickly spread through St. Petersburg of this most unusual fatality, the police were forced to station a guard outside to drive away the curious.

Priestley, writing years later, concluded that Richmann's death was "an enviable one," since electrocution was thought to be painless, but also because the victim had perished in a sort of consummation of his life's passion. Yet the immediate meaning of Richmann's death, discussed among philosophers across Europe, was clearly mixed. It quietly enhanced the idea of Franklin's courage in having "tamed" lightning (obviously, Richmann had not), but also emboldened doubters like Nollet who cited it as an example of why attracting bolts of electricity with a lightning rod was not a good idea, although Nollet neglected to mention that Richmann had been killed experimenting with an apparatus dissimilar from a properly grounded lightning rod. Nollet also seemed to intentionally misconstrue the physics involved by asking why, if pointed rods could protect buildings from lightning, the iron crosses atop many churches could not perform the same function. The answer was that, as these crosses were not grounded, they attracted lightning but offered no way to shepherd it safely to the ground. Franklin addressed this issue in a paragraph he added to an English newspaper account of Richmann's demise that he published in the *Gazette* in 1754, noting that "the new Doctrine of Lightning," i.e., the lightning rod, would spare many such accidents, since had Richmann's apparatus been properly grounded, neither he nor his house would likely have been injured.

Nollet's stature among European natural philosophers guaranteed that his attack on Franklin would be taken seriously, with the result that it probably inhibited the acceptance of lightning rods in

Europe. The abbé renewed his arguments so frequently that he managed to keep the scientific conversation about lightning in turmoil until his death, in 1770, and even beyond. Collinson, who believed Nollet was "farr from Dealing candidly or Ingeniously," told Franklin he suspected that the abbé staged demonstrations in which Franklin's electrical concepts would "fail." Mocking Nollet's reaction to Franklin, Collinson characterized the abbé as one "who look'd on himself as the Prince of Electricians—but on a Sudden Springs up a Little Cloud from the West that Eclipses all his brightness."

Franklin had the least patience for Nollet's fears about warding off the heavenly father's chastening stick. "He speaks as if he thought it Presumption in man to propose guarding himself against the Thunders of Heaven!" Franklin exclaimed to Cadwallader Colden. "Surely the Thunder of Heaven is no more supernatural than the Rain, Hail or Sunshine of Heaven, against the Inconvenience of which we guard by Roofs & Shades without Scruple." Referring to some of his own recent scientific inquiries, he then added, "But I can now ease the Gentleman of this Apprehension; for by some late Experiments I find that it is not Lightning from the Clouds that strikes the Earth, but Lightning from the Earth that strikes the Clouds."

What Franklin had discovered was that, contrary to long-held belief, lightning did not descend from the sky. In a series of observations he made in Philadelphia between April and June 1753, using a system that allowed him to capture electricity from passing storms in a Leyden jar, he determined that thunderclouds are almost always charged negatively. Since, as Franklin had observed in his tabletop experiments, equilibrium is restored when a spark flies from positive to negative, if lightning was simply the atmosphere's way of restoring its own electrical equilibrium, then the same dynamic would hold true in nature. And if clouds were charged for the most part negatively, the sudden "spark" that was lightning must proceed from a positively charged source, either another cloud or the earth. Thus, "for the most part, in thunder-strokes," Franklin informed Collinson, "*'tis the earth that strikes into the clouds, and not the clouds that strike into the earth.*" Franklin had this slightly incorrect: Although he and physicists of his day often referred to the "spark" of lightning, in reality lightning is more in the nature of an electrical connection, not a spark; and the restoration of electrical equilibrium between earth and the clouds entails not solely an upward stroke but a downward one as well, resulting in an oscillation between earth and sky. Yet, his was the first hypothesis to correctly challenge the long-held conception that lightning only strikes downward from the sky to the earth.

In this manner, Franklin gave an even more irrefutable answer to Nollet's theological concerns. If some of the force that created lightning originated in the earth, that meant it was coming from a location opposite to where God presumably dwelled, and there was now even less reason to object to man's "presumption" in trying to defend himself from it. The members of Franklin's Philadelphia Junto, who discussed the issue, concurred, and raised another peculiarity about lightning's "presumption"—its highly questionable aim. "That it is not always directed to execute divine Wrath appears from hence that it most frequently wastes itself on inanimate Things as Trees, Houses, etc." The Junto endorsed the use of Franklin's rods, noting, "So far is it from being Presumption to use this Invention that it appears foolhardiness to neglect it."

Franklin's supporters and biographers are sometimes inclined to treat Nollet as a frustrated also-ran, or even a pest, although in retrospect Nollet's resistance, however motivated, does appear to have played a role in forcing Franklin and other experimenters to further define and clarify their belief in Franklin's ideas. While the lightning rod's capabilities were generally understood, several aspects of its function were legitimately open to debate, and would remain so for years. For instance, it was during this period that Franklin began to relinquish his original idea that the lightning rod would quietly remove electricity from thunderclouds, and accept that its function was to intercept actual lightning strikes and bear them to the ground. Still, Franklin no doubt had the nettlesome Frenchman in mind when he complained in 1768 to Harvard's John Winthrop, "It is perhaps not so extraordinary that unlearned men . . . should still be prejudiced against the use of [metal] conductors, when we see how long even philosophers, men of extensive science and great ingenuity, can hold against the evidence of new knowledge. They continue to bless the new bells and jangle the old ones whenever it thunders. One would think it was now time to try some other trick; and ours is recommended."

Franklin addressed his friend John Winthrop as a familiar ally in the debate over Presumption, for it had been in Winthrop's New England that the lightning rod had recently received its most difficult trial.

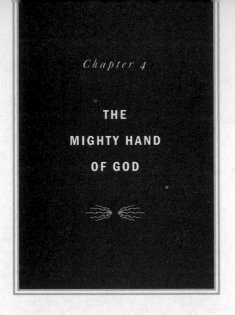

Chapter 4

THE
MIGHTY HAND
OF GOD

In 1755, A MOST SERIOUS ACCUSATION WAS MADE AGAINST LIGHTNING rods: they were said to have caused an earthquake. The powerful tremor that struck southern New England on November 18, and was felt as far away as Nova Scotia and South Carolina, was particularly frightening because it occurred in the middle of the night, at 4:20 A.M., driving Boston residents out of their beds and into the streets in their nightclothes. "Our fears came upon us *suddenly*, WHILE WE WERE SLEEPING," reported one witness, as residents by first light surveyed a scene of collapsed stone walls and broken chimneys, "and . . . many among us . . . were apprehensive for nothing less than the last Judgment, and their Imaginations were so rais'd that they thought they distinctly heard a Trumpet sound."

In the quake's aftermath, a fierce debate ensued over the intertwined issues of providence, electricity, the then-unknown causes of earthquakes, and the clergy's right to speak authoritatively about science, roiling colonial Boston and threatening to darken everywhere the reputation of both Franklin and his invention.

As in earlier decades—there had been a destructive earthquake in Boston in 1727—the 1755 event filled the town's church pews. "God, in his holy Providence, hath been most awfully shaking the Earth," one newspaper announced; said another, "'Tis *Sin*, and that *only*, that enkindles the Anger of Almighty GOD, and causes Him *to march thro' a Land in Indignation*." However, six decades after the collective hallucination of the Salem witch trials, New Englanders had grown more skeptical about the claim that nature was regulated by providence, and few believed anymore that earthquakes expressed God's displeasure or were meant to swallow up the wicked. (Some favored the idea that, although God no longer gave daily attention to the world, he had at Creation preprogrammed natural catastrophes to occur throughout time as a way of reminding humanity of its frailty and the need for constant faith.) This change in sensibility, the waning authority of the New England clerics, and fears about an impending war with the French and Indians (it would start the following year) combined to create a demand for a more rational explanation.

A series of additional tremors across New England in ensuing days did nothing to soothe the population; nor did the news that arrived soon after of an even larger earthquake in the European port of Lisbon on November 1, probably the greatest natural disaster of the century. The crew aboard a ship originating in Lisbon that sailed into Boston Harbor on January 8 told of apartment houses, stacked on top of one another in the hilly Portuguese city, collapsing on their occupants and "the ground thrown up as if by shovels." The Lisbon destruction was vast, with tens of thousands of deaths reported and many killed in a tidal wave that swept the ruined city shortly after the initial quake. Unlike the Jamaican town of Port Royal, a place known for its pirates and debauchery, where the more "understandable" destruction by an earthquake had occurred in 1672, Lisbon was a thriving center of trade, commerce, and the arts—a burgeoning European capital. In the scope of what mankind had come to expect of providence, Lisbon had not deserved its fate. The catastrophe thus stymied not only the efforts of natural philosophers to offer a logical explanation, but even robbed theologians of their usual resolve to link such events to a vengeful God. To ascribe such an act to a supposedly benevolent supreme being was to lay at his feet the killing of thousands of innocents. As the philosopher Susan Neiman has suggested, the Lisbon devastation was the Enlightenment's "Auschwitz," an unprecedented event that profoundly challenged humanity's faith in notions of good and evil. Significantly, the town's leadership, in response to God's action (or inaction) in bringing about such a disaster, sought to limit the church's influence in the recovery, refusing to allow the traditional prayer days, fasts, or pilgrimages that might distract from the business of rebuilding.

Both the Boston and Lisbon earthquakes arrived at a time when some of the first scientific theories about the source of such cataclysms were emerging. "No Doubt natural Causes may be assigned for this Phaenomenon," stated New England minister Mather Byles. The new theories generally suggested agitated subterranean activity involving water or fire. "An imprisond Vapour too closely pent or too strongly compressed in the Caverns beneath," said Byles, "will thro' a natural Elasticity, abhor confinement, dilate and expand, swell and heave up to the surface of the earth, producing a tremor and Commotion."

Cotton Mather had died in 1728, but his penchant for integrating Puritanism with natural philosophy abided in one of his followers, Thomas Prince, the Harvard-educated pastor of Boston's South Church. Like other New England churchmen of the Enlightenment, Prince combined a quasi-erudition on scientific topics, often gleaned from the pages of the *Philosophical Transactions*, with a fundamental assumption of the providential order of nature. As with Increase Mather and others before him, Prince held New England to be one of the Lord's "special places," and he was attuned to the ways in which God continued to act upon and intervene in the region's affairs. His conviction that the divinity spoke through meteorology had been bolstered by a dramatic incident in 1746. Prince was leading prayers before his congregation, attempting to soothe fears over rumors that the French fleet was on its way to ransack Boston, when suddenly the wind loudly rattled the windows. In a moment of inspiration, Prince stretched his arms heavenward and implored God to use the mighty wind to save New England from invasion. Days later, word arrived that a storm had wrecked the approaching French fleet, a devastation so complete that several of its officers had committed suicide. God, it was clear, used the elements not only to punish the wicked, but to defend his "special place."

In the wake of the 1755 tremors, Prince again responded to his anxious congregants, ordering a reprinting of a celebrated sermon he had written after the quake of 1727, "Earthquakes the Works of God and Tokens of His Just Displeasure." The earlier event had brought nearly thirty sermons (such as Joseph Sewall's "The Duty of a People to Stand in Aw of God") from the pulpits of Boston ministers, enlivening the Puritan jeremiad about a society on the receiving end of God's resentment for having abandoned the rigors of its faith. In Prince's 1755 reprint, however, he added something new—the opinion that earthquakes were caused by electricity; and he asserted that Bostonians may have exacerbated the preconditions for a tremor by putting up so many lightning rods, which channeled electricity into the ground. Could it be purely coincidence, he asked, that such a severe earthquake would come just at the time when lightning rods were beginning to appear on public buildings and private homes? The 1753 edition of *Poor Richard's Almanac*, containing the instructions for lightning rods, had sold ten thousand copies in the colonies, a bestseller by any standards of the time, and the apparatus had been well received in Boston, where a huge conflagration in 1747 had painfully reminded citizens of the vulnerability to fire of the town's densely built wooden residences.

"The more Points of Iron are erected round the Earth, to draw the Electrical Substance out of the Air; the more the Earth must needs be charged with it," Prince suggested. "In Boston are more erected than anywhere else in New England; and Boston seems to be more dreadfully shaken. *O! There is no getting out of the mighty Hand of God.*" Lightning rods, he warned, could offer no protection from the "bitter fruits and tokens of His high resentment of the sins of man . . . There is no Safety anywhere, but in his almighty Friendship, and by heartily Repenting of every Sin."

Even as he mimicked Franklin's rational language about points and electrical charges, the minister's grasp of the science was precarious. Prince said that electricity moved about in the clouds in "parties" seeking other electrical "parties," and that clouds were capable of guiding themselves about the sky. While he claimed to accept the need for lightning rods to siphon electricity in its most dangerous form out of the atmosphere, he was troubled by what might occur when electricity, conducted into the ground, joined the substantial amount he believed to be already there, creating harmful "subterranean tension."

Prince's views were not without precedent. Even Franklin, after a mild earthquake disturbed the Middle Atlantic colonies in early December 1737, reprinted in his *Pennsylvania Gazette* an essay from Ephraim Chambers's *Cyclopedia* claiming that "the material cause of thunder, lightning, and earthquakes, is one and the same," a kind of universal fire. In spring 1750, the English countryside had experienced several tremors, and the Royal Society had run a special supplemental edition of the *Philosophical Transactions* solely devoted to earthquakes and featuring no fewer than fifty-seven articles on the subject. One of the British investigators, Reverend William Stukeley, a prominent antiquarian best known for his Christian interpretation of the ruins at Stonehenge, linked the earthquake question to the electricity letters written by Benjamin Franklin that had been read before the Royal Society in 1749. Citing inaccurately from Franklin, Stukeley asserted that the "snap" of electricity in Franklin's experiments, when magnified to the real scale of that "produc'd by a thousand Miles Compass of Clouds, and that reechoed from Cloud to Cloud," would, when directed toward the earth, surely be powerful enough to cause an earthquake. Noting that major earthquakes often occurred near large bodies of water, he surmised that the water might be acting as a kind of colossal Leyden jar, storing and then dispensing a monstrous electrical charge; and he went on to suggest it was more than coincidence that the hysteria, headaches, and aching joints seen in earthquake victims resembled the symptoms felt by people who had inadvertently received an electric shock. Finally, he concluded, these events could only be a sign of God's "chastening rod," since earthquakes almost always struck inhabited towns and cities and rarely "an uninhabited beach" or "bare cliffs."

John Fothergill, in his preface for Franklin's book on electricity, emphasized that Franklin's observations of electrical phenomena were based on actual experimentation, and should not be confused with any quasi-scientific/theological discussion of electricity and earthquakes. Ironically, it was because Franklin's theories *were* so widely accepted that they became fodder in the earthquake debate; after all, if thunder and lightning were caused by electricity, it was an easy leap to the speculation that earthquakes, with their subterranean "thunder," probably were too. The electrical theory of earthquakes also dovetailed nicely with another long-standing superstition—that deep beneath the ground there were pockets of combustible substances poised to explode if agitated. It was with this concern in mind that Franklin's niece in Boston Elizabeth Hubbart sent him "two Specimans of the Sand thrown up by the late Earthquake, if you have a Mind to try any Chymical Experiments."

Only a few miles from Boston's South Church, where Prince preached his sermons, was the classroom of science professor John Winthrop at Harvard College in Cambridge. A descendant of one of the founding families of the Massachusetts Bay Colony, Winthrop personified Harvard's recent move away from traditional religious instruction into aesthetics, language, and experimental science, and his classes were often standing room only. His students may have particularly enjoyed Winthrop's recent notoriety for having suggested, to the chagrin of some Harvard elders, that dancing was the human expression that most closely mimed nature "in its highest perfection." Since 1746 he had been giving lectures on electricity, and had performed some of Franklin's experiments for his students, using equipment Franklin helped the school obtain.

The earthquake had badly shaken the buildings in Harvard Yard, so Winthrop, like Prince, was only reacting in his professional capacity when he went before his "congregation"—a hall packed with curious students—to try to explain what had occurred. Winthrop said that while he lacked conclusive proof, he believed earthquakes were caused by a "kind of undulatory motion," which he compared to a wave, that moved both horizontally and vertically beneath the earth's surface. He related that during the recent quake he had seen buildings near his home sway back and forth in a pendulum-like action. The bricks from his chimney, which was

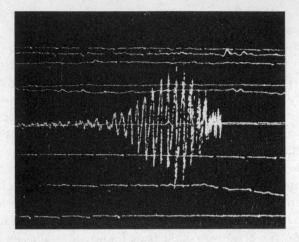

thirty-two feet high, had fallen thirty feet away; thus, he calculated the speed of the tremor at twenty-one feet per second, and noted that the shorter in duration the earth's vibrations were, the quicker the pendulum effect. Winthrop's ingenious measurement of the distance that bricks had been thrown from collapsing roofs and houses—surely some of the world's first computations of earthquake science, or seismology—led him to state that "our buildings were rocked with a kind of angular motion, like that of a cradle; the upper parts of them moving swifter, or thro' greater spaces in the same time, than the lower; the natural consequence of an undulatory motion of the earth."

Prince's published sermon warning that lightning rods caused earthquakes and Winthrop's offering of prescient scientific hypotheses proved to be the opening salvos in a yearlong public feud. Winthrop disputed Prince's claim that the earthquake had "targeted" Boston, pointing out that it had been reported elsewhere and that Boston suffered more damage simply because it had so many buildings built of brick. He added that underground there were not known to be large entities, such as clouds, capable of being charged electrically, nor evidence of any subterranean electrical system that would function in the way Franklin had described the action of lightning. And echoing Franklin's reply to Nollet, Winthrop took on Prince's charge of presumption, writing in defense of the lightning rod in a letter to the *Boston Gazette:* "It is as much our duty to secure ourselves against the effects of lightning as against those of rain, snow, and wind, by the means God has put in our hands."

Prince began his reply, also published in the *Gazette,* evenly enough, declaring that no real argument existed between him and Winthrop. As a clergyman, he explained, it was his task to concern himself with the ways of God, while as a professor of science, it was Winthrop's to ponder the mysteries of natural phenomena. But Prince also suggested that Winthrop, as a younger man, needed to show greater deference to someone, like himself, who had survived the 1727 earthquake, and whose views on earthquakes were so widely accepted. Winthrop replied that they could not possibly be in agreement if Prince insisted that lightning rods caused earthquakes. He conceded there was an old folk belief that lightning and earthquakes were related because each sometimes produced a sulfurous odor, but emphasized that that similarity alone could hardly support blaming the quakes on lightning rods.

Winthrop probably offended Prince most not by questioning God's agency in earthquakes but by casting doubt on the cleric's credentials as a natural philosopher. In a sense, Winthrop's objection to Prince's science was the same that William Douglass had raised against Cotton Mather during the inoculation battle of 1721, that church leaders, however large their congregations, not be allowed the automatic privilege of extending their authority to the empirical world of science. Winthrop's criticism highlighted a growing deficiency in the New England clergy's recent tradition of mixing a little Newton with their theology. As humanity's knowledge of natural science became more complex, it was only a matter of time before those who sought to broker a place for God's agency in natural phenomena would be revealed to have a faulty comprehension of the subject.

"I cannot but esteem it a high felicity to have rescued this worthy Divine from the panic which has seized him," wrote Winthrop, upbraiding Prince for slandering lightning rods and needlessly scaring people. "Consequentially, a great number of others, especially of the more timorous sex . . . have been thrown into unreasonable terrors, by means of a too slender acquaintance with the laws of electricity." Prince's declarations, Winthrop feared, would "discourage the use of the iron-points . . . which, by the blessing of God, might be a means of preventing many of those mischievous and sorrowful accidents, which we have so often seen to follow upon thunderstorms."

He saved his strongest criticism for Prince's remarks about presumption, his *"O! There is no getting out of the mighty Hand of God,"* which Winthrop viewed as a desperate appeal by a fading Puritan to revive the spirit of the jeremiad: "[I] cannot believe, that in the whole town of Boston, where so many iron points are erected, there is so much as one person, who is so weak, so ignorant, so foolish, or, to say all in one word, so atheistical, as ever to have entertained a single thought, that it is possible, by the help of a few yards of wire, to 'get out of the mighty hand of God.' "

The professor had not only exposed Prince's weakness as a natural philosopher, but also outflanked him on theology. While from the twenty-first century their dispute appears to be one of superstition (Prince) vs. rationality (Winthrop), in its eighteenth-century context, it was a far more nuanced quarrel about God's relationship to nature. God was central to the arguments made by both sides. Prince believed in an all-controlling divinity; Winthrop, far from denying a divine role in human affairs, held that God's plan for humanity was itself rational: Man, in discovering the workings of the cosmos, was only doing what God intended, and he would surely approve of humanity's efforts to thwart the dangers of lightning. Franklin, after all, had introduced the lightning rod to the world in *Poor Richard's* with the words "It has pleased God in His goodness to mankind at length to discover to them the means of securing their habitations and other buildings from mischief by thunder and lightning." The most pious believer, it seemed to Franklin and Winthrop, could employ a useful item like a lightning rod without fear it would in any way alter his relationship with God.

Prince ultimately withdrew from the earthquake dispute with Winthrop, covering his retreat by remarking that people surely had better things to do than concern themselves with the *science* of earthquakes when they should be reflecting on God's unmistakable warning to the people of Boston. He couldn't resist a final effort to belittle Winthrop, however, suggesting that the Massachusetts Assembly, out of deference to "Mr. Professor's" famous ancestors, grant the teacher a hundred pounds a year for scientific study.

One person who took a great interest in the Winthrop-Prince controversy was a former student of Winthrop's, John Adams, the future president of the United States, who, when the earthquake struck on November 18, had been at his father's house in Braintree, near Boston. Adams cherished a memory of Professor Winthrop leading him and several other students onto a rooftop in Harvard Yard one clear night to peer through a telescope at the moons of Jupiter. While he admired his former teacher's retort about no one being so foolish as to think lightning rods removed them from the mighty hand of God, he observed in his diary that, in fact, more Boston residents were inclined to concur with Prince's warning. Adams, who suspected religion of having inculcated in mankind an inordinate fear of nature, complained that he had heard persons "of the highest rank" decry the Franklin rods as "an impious Attempt to robb the Almighty of his Thunder, to wrest the Bolt of Vengeance out of his Hand." In the end, Adams's assessment of the Winthrop-Prince affair proved accurate. Although to the literate onlooker Winthrop had thoroughly vanquished Prince, the Puritan minister had, it seemed, more effectively swayed public opinion, and it was said that the use of lightning rods in Boston declined for many years thereafter.

Franklin's fame was associated with his discoveries in electricity, but he never left off his diligent observation of nature's subtleties in all its forms, or his tinkering with potentially useful inventions. His correspondence, even during his diplomatic missions abroad, is full of detailed accounts of inquiries into a range of subjects, as well as thoughtful and often probing responses to other natural philosophers of his wide acquaintance.

In 1757 he traveled to England as agent of the Pennsylvania Assembly to negotiate with the Proprietors of Pennsylvania, Wil-

liam Penn's sons Richard and Thomas, and with the British government, over relations between the Proprietors and the citizens of
Pennsylvania. The major outstanding issue between the Assembly
and the Penns had to do with the latter's unwillingness to be taxed
on their huge property holdings in the colony. Thomas Penn had a
wary respect for Franklin, based largely, of all things, on Franklin's
role in the "Plain Truth" episode, in which Franklin had helped
safeguard the Penns' colony from possible attacks by privateers in
the lower Delaware River. Penn, noting Franklin's skill as a rabble-
rouser, expressed concern that Franklin, or any individual, should
"be always in the heads of the Wild unthinking Multitude." He is,
said Penn, "a dangerous Man and I should be very Glad he inhabited any other Country, as I believe him of a very uneasy Spirit.
However as he is a Sort of Tribune of the People, he must be treated
with regard."

In England, Franklin was received warmly by the country's
leading men of science, many of whom he had corresponded with
for years. In 1760, having been made a Fellow of the Royal Society
for his electrical studies, he took a seat on its council. Franklin had
the pleasure of finally meeting Peter Collinson face-to-face, and
he also got to know another longtime correspondent, publisher
William Strahan, who had arranged for the printer David Hall to re-
locate to Philadelphia and assume Franklin's duties at the print
shop. Collinson and Strahan were both smitten with "the fat old
fellow," as Franklin had described himself in a letter notifying
them of his arrival. (To the amusement of his London landlady
Margaret Stevenson and her daughter Polly, he sometimes called
himself "Dr. Fatsides.") "For my own part," Strahan confessed in
a note to Deborah Franklin, "I never saw a man who was, in every
respect, so perfectly agreeable to me. Some are amiable in one
view, some in another, he in all . . ." (Unfortunately, Franklin and
Strahan's great friendship, built on their common interests in science and literature, would not survive the coming politics of the
American Revolution.) Franklin and his son, William, who had accompanied him to study law, would remain in England for the next
six years.

A far cry from his original visit to London in the 1720s as a struggling young journeyman printer, this trip found Franklin returning as a respected scientist, publisher, and statesman. Welcomed in the parlors of leading politicians and natural philosophers in this relatively stable time in British-Colonial relations, before the imposition of the restrictive Stamp and Townsend Acts, he could enjoy the stimulating life of the gentleman scholar. So taken was he with life in England, he even considered relocating his Philadelphia family there.

His snug happiness with his new world was epitomized in an invention he completed within a couple of years of his arrival, a homespun musical instrument in which glass disks were turned by a treadle and rubbed gently with the fingers, the *armonica*. It was perhaps the least practical invention Franklin ever made, but one he always claimed to be his favorite.

The 1750s were a time of mild upheaval in the world of music, as the patronage system, so long in practice for composers and musicians, entered its decline and the ideal of the independent musical virtuoso began to replace it. The era produced a new expressive keyboard instrument, the piano, as well as standardized musical notation that made compositions available to even amateur musicians. One of the leaders of this movement away from formality in music was a contemporary of Franklin's, George Frideric Handel, the German composer who had adopted England as his home in 1712 and had in return become one of its most celebrated artists. His large secular works, using folk melodies and even the cries of street vendors, brought classical music to the people with an accompanying element of spectacle.

Franklin, who had very strong ideas about music, also preferred folk ballads, as well as a natural method of singing, and strong melody and harmony. He worried that "modern" music increasingly encouraged a kind of empty virtuosity, and that audiences listening to rapidly played passages were experiencing only "the same kind [of] Pleasure we feel on seeing the surprizing Feats of Tumblers and Rope Dancers, who execute difficult Things." He disapproved of vocal acrobatics such as forced drawling, stuttering, the placing of emphasis on words of no importance, and "screaming without cause." While Franklin had a good comprehension of musical theory, as a musician he was probably never more than a dedicated amateur. The one musical composition with which he's credited, the "Quartet in F Major" (also known as "The Open String Quartet"), seems more an experiment meant to show that a great amount of effort might be saved in the playing of the violin, for he had observed that the instrument could essentially be played with one hand if it was tuned to an open set of notes, thus enabling the performer to concentrate exclusively on bowing.

In May 1759 Franklin was in the audience at a benefit performance of Handel's *Messiah* (a fortnight after the composer's death). A short time later he learned of a man named Richard Pockridge who performed Handel's *Water Music* and other classics on a set of tuned wineglasses he called the "Angelic Organ." Rubbing the rims of the glasses with his moistened fingers, Pockridge produced an ethereal sound that had a distinct, soothing quality. Pockridge—"Old Pock," as he was known—was an eccentric and a dreamer: decades before they would become reality, he advocated schemes for "unsinkable" ships, human flight, and blood transfusions to reinvigorate the elderly. Tragically, however, he, his visionary ideas, and the Angelic Organ succumbed in November 1759 in a blazing London house fire. The first set of musical glasses Franklin heard belonged to Edmund Hussey Delaval, who had made a copy of Pockridge's instrument and demonstrated it before members of the Royal Society.

"Being charmed by the sweetness of its tones and the music he produced from it," Franklin said, "I wished only to see the glasses disposed in a more convenient form, and brought together in a narrower compass, so as to admit of a greater number of tunes and all within reach of hand to a person sitting before the instrument."

The musical method of rubbing fingers on the rims of glasses or bowls filled with water first appeared in Europe in the late Middle Ages (Galileo, himself the son of a musician, experimented with it), although sets of tuned glasses—"tuned" by the amount of liquid they contained—didn't receive serious attention before 1746, when Christoph Gluck performed a "Concert of Musick . . . upon Twenty-six Drinking Glasses, tuned with Spring water," at the Little Theatre in Haymarket. Not until Franklin, however, had anyone conceived of turning them into a stand-alone musical instrument.

Working with a London glassblower, Franklin began by doing away with the drinking glasses, substituting a series of thirty-six glass disks constructed to a specific thickness and size. This design was no doubt inspired by the electrical friction machines he knew so well that rubbed a sympathetic material against a rotating glass ball, as well as by his own "electric battery" that used plates of glass to replicate the storage capacity of several Leyden jars. The disks were gathered on an iron spindle and arranged in diminishing size from left to right, each separated from the other by a cork buffer. This way, the performer no longer needed to wipe his finger gently around an upright glass, but could simply allow the spindle to turn the disks—or "cups," as they were called—under his finger, which he kept moistened with a damp sponge. The cups, representing three octaves extending from a low G, were painted different colors to represent the tones and semitones. The instrument emitted a gentle keening tone, the sound of each individual cup sustaining and combining with others just played. "The advantages of this instrument," Franklin wrote, "are that its tones are incomparably sweet beyond those of any other; that they may be swelled and softened at pleasure by stronger or weaker pressures of the finger, and continued at any length; and that the instrument, being once well tuned, never again wants tuning." The contraption rested in a wooden case and stood on four legs, and was turned either by hand or, in many models, by pressing on a foot treadle.

Thousands of armonicas were built and sold—they could be ordered as either "a Portable Instrument, or [a] Genteel Piece of Furniture"—and many composers wrote music for them. Two armonica virtuosi emerged. Marianne Davies, who may have been taught to play by Franklin, and who was probably given an instrument by him, toured Europe successfully, accompanied by her sister Cecilia, a singer. During a stay in Paris, Davies is said to have taught Marie Antoinette how to play. An even more famous star of the armonica was Marianne Kirchgessner, a blind woman whose performance inspired one member of her audience, Wolfgang Amadeus Mozart, to write an armonica composition in her honor, the

"Adagio and Rondo for Glass Harmonica, Flute, Oboe, Viola and Cello"; Kirchgessner and Mozart performed the piece together in Vienna, with Mozart on viola.

Franklin always claimed that the armonica was the invention he took most delight in, and certainly he seemed at times almost giddy with it, keeping one with him wherever he happened to be living, and often demonstrating it at parties, much as he'd once given "entertainments" of electrostatics. Word of Franklin's musical recreations amused those friends who learned of them back in the colonies; Thomas Penn was heard to complain that Franklin was foolishly wasting his time on "philosophical matters and musical performances on glasses." But Franklin's devotion to the instrument remained steadfast. At home in Philadelphia in late 1762, he installed an armonica in his third-floor music room, which also held his daughter's harpsichord along with various bells and other instruments. He and Sally played duets: some classical pieces, but mostly the Scottish folk tunes Franklin liked. Many of his Philadelphia friends heard him perform on the instrument, and there is an irresistible anecdote that Deborah, awakening one morning to the sound of him practicing, concluded she had died during the night and gone to heaven. On another occasion Franklin was said to have played an impromptu recital for the mother of a girl his son William had recently jilted, turning a potentially uncomfortable visit into "a very easy afternoon."

The vogue for the instrument proved to be of relatively brief duration. The armonica's soft, subtle tones couldn't compete with the piano, which was capable of greater volume and which better complemented the bigger orchestras filling stages in ever larger concert halls, and it soon also wore out its welcome in the drawing room. Concern had spread that the high lead content of the crystal used in armonica disks and the constant rubbing of one's fingers in a fixed position caused mysterious neurological symptoms. There also arose fears that even listening to the armonica was dangerous— "an apt method for slow self-annihilation." Its haunting tone and deep sustain did have a numbing effect on listeners, so much so that it was later used by Franz Mesmer and other healers to put patients into a trance. Tales of it inducing convulsions in cats and dogs, causing women to give birth prematurely, and even waking the dead, were published, and though Franklin himself played the instrument for years without apparent harm, both of its leading virtuosi, Marianne Davies and Marianne Kirchgessner, experienced nervous disorders, with Davies forced to spend more than a year recuperating alone in a room.

During his second period of residency in England (1764–1775), this time representing not only the interests of Pennsylvania but also those of Massachusetts, Georgia, and New Jersey, Franklin was involved in a scientific debate that was one of the most stimulating of the Enlightenment: the question of the age of the earth and of living things, including man. Like the arguments about the lightning rod's "presumption," this inquiry challenged long-received ideas about the relationship between God and man, and went so far as to call into question the biblical version of Genesis and Creation.

Franklin's involvement began in 1764 when, at a salt marsh on the Ohio River about forty miles south of present-day Cincinnati, an Indian agent and land speculator named George Croghan discovered extensive fossil remains of about thirty large prehistoric mammals. "This Mud being of a salt quality is greedily lick'd by Buffalo, Elk & Deer, who came from distant parts, in great Numbers for this Purpose," Croghan noted, describing the location that would come to be known as Big Bone Lick. Soon after, his party endured an Indian skirmish in which Croghan lost five men and he "got the stroke of a Hatchet on the Head." He had no choice but to abandon the fossils as he and other survivors fled for their lives. A year later he returned, gathered more bones, and this time succeeded in shipping some of the relics to contacts in London, including Franklin, whom he knew through his involvement in Pennsylvania's frontier affairs. Franklin received from Croghan four huge tusks, a segment of vertebrae, and several enormous teeth. Naturalist James Wright, who saw the Croghan relics while they were still in America, assured John Bartram that "the Creature when Alive must have been the Size of a Small house," and passed along the Indian myth that the heavenly father, to protect the local Indians, had used lightning to strike the animals dead.

In Franklin's time, the study of the earth's oldest living things, later known as paleontology, was just emerging as an area of scientific inquiry. Philosophers had seemingly put off indulging this especially ticklish speculation as long as possible, fully aware it dared condemnation by church or state. But it was a logical outgrowth of the craze for Linnaean classification and the gathering of specimens like snake skins and tortoise shells that had gripped the colonies for several years.

This new curiosity raised unique, far-reaching questions. What was the true age of the earth and of its living things? Were some of God's creatures that had once lived now extinct? What could God have meant by such an action? Of these and other delicate issues, extinction was the most disturbing, upsetting not only biblical tradition but even the Newtonian universe admired by the *philosophes* which, while not tended by God, had certainly been created by him as a smoothly functioning mechanism. Why would a divinely inspired machine have parts that ceased to belong, and that ultimately disappeared?

Suspicions about a remote past that predated biblical time, and the notion of extinction, had long prowled at the fringes of learned thought. For centuries it was believed that humans were descended from a race of giants. Bones found in a quarry in southeastern France in the 1600s were held to be those of Teutobochus, a legendary thirty-foot-tall Teutonic king who led armies against the invading Romans and perished before the birth of Christ; the "remains" of this warrior were displayed in marketplaces and at fairs across Europe. Other unearthed teeth or bones of "giants" were often displayed by churches as relics of saints, and both the cyclops and the unicorn, mythical creatures, were born from man's uncertain conclusions about fossil remains. The cyclops's solitary eye was suggested by the gaping proboscis cavity of extinct dwarf elephants; the unicorn legend arose from the fossilized tusks of elephants and rhinoceroses, which, prized for their magical and medicinal virtues, were traded both by the ancients and in medieval Europe.

The remains that Franklin's friend George Croghan had found at Big Bone Lick were those of mastodons, elephant-like creatures with heavy coats and huge upward-curving tusks that are said to have appeared anywhere between about 20 million and 3.5 million years ago, and survived until as recently as 10,000 years ago. Related early fossil finds were those of a smaller animal of the

order Proboscidea, the mammoth, thought to have emerged about 2 million years ago.

In the eighteenth century, of course, little of this was known. In 1712, when Cotton Mather reported to the Royal Society that a tooth weighing more than four pounds and a thigh bone seventeen feet in length had been unearthed near Albany, New York, he assumed—as did a number of his contemporaries—that the remains were those of a giant man who had perished in the Great Flood. African slaves in America were likely the first to point out that the bones unearthed at sites in New York and Virginia resembled those of the elephant.

Franklin studied the relics Croghan had sent and replied that it was his belief that the Ohio bones, as well as recent finds of similar bones in Siberia, all came from the same animal, a kind of carnivorous elephant. Franklin, who called the Ohio discovery site "The Great Licking Place," speculated that what the find indicated was that the climates of the earth may have once been inverted, allowing elephants to roam far to the north of their known habitats in Africa and the Indian subcontinent. In a second letter on the subject, written in 1768 to a French acquaintance, the astronomer Jean Chappe d'Auteroche, Franklin reversed himself on the carnivorous aspect, saying that his study of the teeth and tusks had led him to believe the animal was "too bulky to have the Activity necessary for pursuing and taking Prey." The find at Big Bone Lick, and recent reports of the discovery of sea fossils in mountain altitudes, prompted Franklin to muse that the earth had a deeper geophysical history than he heretofore had imagined. "'Tis certainly the Wreck of a World we live on!" he wrote, praising as beneficial "those convulsions which all naturalists agree this globe has suffered" for having exposed various strata of "clay, gravel, marble, coals, limestone, sand, minerals, etc." Ever the pragmatist, he wondered if the eons during which the earth had apparently experienced great geological readjustment had been a "means of rendering [it] more fit for use, more capable of being to mankind a convenient and comfortable habitation."

Franklin's friend Peter Collinson, who had also been informed by Croghan about the bones found along the Ohio River, made the important suggestion that while the creature might be related to the elephant, it was not the same animal, and that it may exist only in fossils. Reluctant to suggest extinction, however, he speculated that the remains might be those of some kind of elephant that, living in southern climes at the time of the biblical deluge, had been drowned in the Great Flood and swept northward. William Hunter, another man of science who studied the mastodon remains shipped to London, was bolder, concluding that they had belonged to a large carnivorous animal—he called it "*Ohio incognitum*" or "the American incognitum"—and that it had vanished long ago. He noted that "though we may as philosophers regret [its disappearance], as men we cannot but thank Heaven that its whole generation is probably extinct."

Extinction, then as now, was a heavy concept, and to pursue such a theory brought one square against not only prevailing views of God's kingdom but the accepted wisdom about the age of the earth itself. In the early seventeenth century, James Ussher, an Irish archbishop and vice-chancellor of Trinity College, had, by calculating backward the life spans of the patriarchs, established the creation of the world as having taken place at 6 P.M. on October 22, 4004 B.C. Ussher died in 1656, honored as Britain's most distinguished biblical authority, and his widely respected chronology—his "fantastically precise misconception," as H. G. Wells termed it—had by 1775 long been a fixture in the margins of most popularly sold editions of the Bible. Franklin himself had published in *Poor Richard's* some excerpts from a popular chronology of the history of commerce that dated the Flood at 2348 B.C., although Franklin likely reprinted the material chiefly for its comical fastidiousness about a number of pseudo-momentous dates in human history, such as the invention of playing cards (1391) and the first silk stockings worn by a king (1547).

Franklin had closely followed events in the 1730s surrounding one of the first scientific developments to cast doubt on the biblical explanation of existence, the discovery by biologists of the polyp, a tiny flower-like water organism that possessed the power of regeneration. Slice the polyp in two, it became two polyps; cut off a tiny part, it grew a replacement. Later this peculiarity was

seen in worms, starfish, and lobsters. These finds sparked a trend of interest in the polyp and a flurry of articles about regeneration, but drew gasps from the clergy, who had no immediate explanation of how the soul, thought to reside in all living things, could be accounted for under such circumstances. The polyp's emergence gave new emphasis to the mechanistic explanation of animal and human life associated with René Descartes, an explanation that had always been at sharp odds with church doctrine. The polyp had a lasting impact on natural philosophers, and its regenerating capabilities often found their way into Franklin's political analogies, inspiring what may have been America's earliest political cartoon, a snake cut up into segments labeled for the American colonies. Drawn by Franklin for his *Pennsylvania Gazette* in 1754, at a time when he was playing a leading role in plans for colonial union, it bore the caption "Join or Die."

Ever willing to be the occasional gadfly in the institutionalized world of European science, it was the Comte de Buffon in Paris who took some of the first steps toward challenging standard notions about the world's age. His prominence and his lengthy sinecure at the Jardin du Roi owed a great deal to his care in skirting controversy with Versailles or the church. As it was risky for any philosopher to dare describe the workings of the natural world without some reference to theology, Buffon in his early science writing had conscientiously incorporated sources from antiquity or the Bible. "It is better to be humbled than hanged," he resolved after a close call in 1749, when authorities forced him to publish a retraction to his suggestion in *Histoire Naturelle* that the earth and the other planets were formed by a collision between a comet and the sun. In 1778, however, now looking toward his legacy and determined to leave behind a thesis of Newtonian scale, Buffon sum-

JOIN, or DIE.

moned the courage to formally dispute biblical interpretation. In a foundry he owned, Buffon experimented with heated metal globes and the time they took to cool, measuring, by analogy, the original molten heat of the earth and the years required for the earth's surface to cool adequately to support human life. In a new volume of his *Histoire Naturelle*, which he titled *Les Époques de la Nature*, he identified seven stages in the earth's history (perhaps an attempt to assuage Christian sensibility with its correlation to the seven days of Creation), and announced that the world was 74,832 years old. This estimate, while a body blow to tradition, was even in Buffon's mind conservative, for in an unpublished manuscript copy of *Époques* he had guessed that the age of the earth might be as much as 3 million years.

Buffon also spoke in favor of a theory of extinction. Having examined mastodon bones sent from Big Bone Lick, he concluded that "every thing leads us to believe that this ancient species, which must be regarded as the first and largest of terrestrial animals, has not existed since the earliest times." The mastodon's extinction seemed evident since it no longer could be found anywhere, he argued, for "an animal whose species was larger than that of the elephant, could hide itself in no part of the earth so as to remain unknown."

One person in vehement disagreement with the latter contention was Thomas Jefferson, governor of Virginia from 1779 to 1781. Jefferson took pride that so important an archaeological find as *Ohio incognitum* had occurred in his young country, and in his

Notes on the State of Virginia (1781) he cataloged the mastodon as one of the United States' living creatures and insisted it roamed still in North America's vast wilderness. In so doing, he rejected extinction. "Such is the economy of nature," Jefferson wrote, "that no instance can be produced of her having permitted any one race of her animals to become extinct, of her having formed any link in her great work so weak as to be broken."

Jefferson also took issue with Buffon and the abbé Guillaume Raynal, author of an influential work on the New World, *A Philosophical and Political History of the Settlements and Trade of the Europeans in the East and West Indies* (1770), because of their claim that animals and men degenerated in the unsavory environment of the New World, and as a result were physically smaller than their European counterparts. Jefferson responded to Buffon and Raynal by reminding them in *Notes on the State of Virginia* that *Ohio incognitum* was "the largest of all terrestrial beings" and dwarfed anything Europe could offer. Jefferson, known to have a passion for measurement, included a chart in which he listed a series of animals by their French and English names and detailed their vital statistics. Beaver found in Europe weighed on the average 18.5 pounds, while the American version weighed as much as 45 pounds; the flying squirrels of Kentucky were twice the size of those in Belgium; and then there were those creatures unique to North America that the Europeans could not claim at all—the magnificent buffalo and the wild cat (*chat sauvage*). To help make his point, he commissioned a hunting expedition to go to New Hampshire and obtain the skin and skeleton of a moose, which he had crated and shipped to Buffon at the Jardin du Roi. Later, as president of the United States, one of Jefferson's first acts upon the return of Meriwether Lewis and William Clark from exploring the Louisiana Territory in 1806 was to dispatch Clark to Big Bone Lick to collect additional relics, which he then stored in the East Room of the White House.

Franklin fought the same battle as Jefferson, but with his customary levity. Seated at a dinner party in Paris at which Abbé Raynal was discussing his and Buffon's degeneracy theory, Franklin interrupted to suggest that the issue might be easily settled if the French and Americans attending would rise from their chairs. When everyone stood up, the Americans towered over the French;

III. *Domefticated in both.*

	Europe.	America.
	lb.	lb.
Cow	763.	*2500
Horfe		*1366
Afs		
Hog		*1200
Sheep		*125
Goat		*80
Dog	67.6	
Cat	7.	

Raynal himself appeared especially puny. "In fact," one guest later recounted, "there was not one American present who could not have tost out of the Windows any one or perhaps two of the rest of the Company."

In these subjects—regeneration, the age of the earth, extinction, degeneracy—natural philosophy felt its way into profound areas that for centuries had been essentially off-limits to secular inquiry. The discussion invigorated by Croghan's discovery at Big Bone Lick, the critical observations of Franklin, Collinson, and others, and Buffon's pronouncement in *Époques*, helped set the stage for the vast expansion of the discipline in the nineteenth century, culminating in the work of Georges Cuvier, who would lay the formal groundwork for paleontology, and Charles Darwin's theories of evolution and natural selection. America's first museum of fossils and paleontological curiosities, including mastodon relics, would be that operated by the Philadelphia artist Charles Willson Peale, in conjunction with the American Philosophical Society.

As Franklin's lightning rod in New England suffered the calumny that it contributed to earthquakes, it began to win acceptance in

Europe thanks to another recent technological development: the increased size and use of artillery in regimental armies, and the related need to store large amounts of gunpowder. Vaults beneath sturdily built churches had traditionally served this purpose, but it was found that when churches full of ammunition were struck by lightning, the instantaneous results were catastrophic. Accidents of this kind caused some of the largest "man-made" explosions the world had ever known, the most infamous being when lightning struck the church of San Nazaro at Brescia, in the Republic of Venice, on August 16, 1769, blasting the church to fragments, leveling much of the town, and killing an estimated three thousand inhabitants. In a magazine underneath the church, the Republic had stored 207,600 pounds of gunpowder. A dispatch from Italy published in the *London Chronicle* explained:

> Since we had the misfortune of having our powder magazine blown up, three days have been spent in digging out of the ruins about two thousand bodies, and five hundred people, who though they are still alive, are dangerously bruised. The fifth part of this city is entirely destroyed, and the rest is very much damaged. Orders have been sent from Venice to leave off searching for the dead bodies, as the corruption of so many might affect the air, and occasion several distempers. The damages suffered by this deplorable accident are estimated at two millions of ducats.

According to eyewitnesses, the initial explosion launched the Brescia tower into the air in one piece; it then fell to earth "as a shower of stones" and hurled fragments so far over the town and surrounding countryside that investigating officials later admitted they hesitated to specify the exact distance for fear no one would believe them. "The effect surpassed anything which has ever been, either before or since, deplored by men," noted French natural historian W. De Fonvielle, who could only conclude, "An observing mind remains in anxious suspense before the marvels of electricity."

The Brescia disaster helped foment a bitter controversy in

England over Franklin's invention. In 1771–72, Parliament passed laws authorizing government supervision of gunpowder storage and manufacture, including the large magazine at the new Royal Arsenal at Purfleet on the Thames, where British men-of-war regularly docked to take on supplies of ammunition. The government's Board of Ordnance asked the Royal Society to investigate and recommend what kind of lightning protection would be appropriate, and the Society in turn appointed a committee of experts to consider the issue. These included Benjamin Franklin; Benjamin Wilson, a portrait artist and natural philosopher; and the venerable English electrician William Watson.

Franklin probably enjoyed the deference of his fellow committee members, for lightning rods were already being commended in Europe for bringing long-needed protection to several prominent edifices. St. Mark's Basilica in Venice, its steeple reaching 340 feet into the sky, had historically been one of lightning's favorite targets, suffering fire damage and loss of life in 1388, 1417, 1489, 1548, 1565, 1653, 1745, 1761, and 1762. After a lightning rod was installed in 1766, the famous tower had no more destructive visits by lightning. Voltaire had a lightning rod placed atop his house at Ferney, although his former patron Frederick the Great forbade any rods to be placed atop his Potsdam abode, Sans Souci. The proud Frederick, according to one history, was upset that "an unpretending naïve man, a species of clodhopper, had discovered a great secret, which had escaped all the physicists of his court, all the powdered, scented, frilled professors who sat in the Royal Society of Berlin." The Prussian Ministry, however, disregarded Frederick's bias and put lightning rods on its roof.

The work of the Purfleet committee involved the close consideration of how lightning acted upon various types of structures, the kind of analysis Franklin had been reporting anecdotally in his *Gazette* for years. Lately, such detailed accounts of lightning strikes had become something of a cottage industry. When a chapel on Tottenham Court Road in London was struck in early 1772, for example, the result was an eight-page report to the Royal Society diagramming the lightning's progress through the building. Hitting "an ornament representing a pineapple carved in wood" that rested atop the structure, the incoming charge then diverged: one splinter found its way to the lead casement of a ground floor window; the other, attracted to a small bell above the front door, zoomed downward to melt the parts of a clock, and then entered the lining of the door itself. Almost incidental to the report's meticulous account of electricity's travels through the building is the information that a lone man who seemed to have wandered in between services to worship privately, one "Goodson, aged thirty-four, by trade a taylor," had been killed. "The poor man destroyed by this accident was sitting at the time on a short ladder, which lay horizontally on the pavement, with his back against the door. The lightning flew from the middle bolt, and struck him on and under his left ear, entered his neck, making a wound half an inch long, passed down his back, which it turned black as ink, down his left arm, melting the stud in his shirt sleeve." The corpse was put on a table where, after two days, it was observed not to differ substantially from "bodies . . . which die a natural death."

Another account in the *Philosophical Transactions* recorded lightning's effects on an elegant home at Naples, Italy. On March 15, 1773, a Lord Tylney hosted a large party to which he invited "most of the nobility of this country . . . [and] foreigners of distinction." When a storm burst overhead and lightning struck the crowded mansion, panic ensued. A Polish prince who had been playing cards leapt to his feet, thinking some sort of attack was occurring, and "clapping his hand to his sword, put himself in a posture of defense." One servant, "asleep on the . . . staircase, his head reclining against the wall, had the hair entirely singed from it on that side." An examination the next day revealed that all the gilt trimming in the house as well as the bell-wires had been burnt, and that the worst damage occurred where the lightning encountered "interruptions in metallic continuity."

When the Purfleet committee visited the Royal Arsenal, Franklin and the others found five rectangular brick buildings with arched roofs and several rooms stacked with barrels of gunpowder. The barrels, rimmed with copper, rested on long wooden grooves supported by bars of iron. Immediately they saw that the buildings were filled with "broken conductors"—metal components sufficient to conduct an electrical charge into the structure but not to lead it safely out. In all lightning rod constructions, Franklin recommended a connection for grounding purposes into soft, moist earth; at Purfleet, because of the devastating potential consequences of inadequate protection, the committee suggested that several wells be dug around the buildings. The wells would support a lead pipe that, at ground level, would connect with an iron bar, flattened to the wall, "and extending ten feet above the ridge of the building, tapering from the ridge upwards to a sharp point."

The committee members favored the installation of pointed lightning rods on the arsenal, although Wilson held to the view that points draw down lightning that otherwise might pass harmlessly overhead. "By points we solicit the lightning," he said, "and may promote the mischief by drawing the charges from charged clouds, which would not discharge at all on the building if there were no points on the conductors." He claimed that Franklin's points would draw electricity from twelve hundred yards away, while the blunt rod he recommended would do so only if electrified clouds came within one hundred yards. As Wilson had written in 1764 after lightning brought "great injury" to St. Bride's Church in London, "A strong electricity breaks through everything we can oppose to it, and unfortunately, thunder is the greatest of electricities." Points, he worried, had a dangerous "power of invitation."

Wilson's thinking about blunt rods may have been influenced by an experiment conducted by the physicist Beccaria. In 1753 Beccaria, a leading Franklinist, erected on a church rooftop in Turin an iron rod capable of being manipulated from the ground so that its pointed end faced, alternately, toward the sky and toward the earth. When the atmosphere was electrified, sparks appeared if the rod was faced upward, toward the sky, but not when Beccaria turned it to face downward. The experiment reinforced what Franklin had said of points—that they would readily draw electricity—but also seemed to underscore Wilson's concerns.

The problem that Wilson saw, and that Franklin and his supporters ultimately had to admit, was that the pointed rod seemed not so much to quietly draw away a charge, but to serve as a contact between earth and sky when lightning struck. Wilson wondered why, if the pointed rod was not going to drain off electricity adequately, anyone would want one on their roof, attracting lightning. Franklin's response to this question was that, even if the rod failed to draw off the charge, it would nonetheless conduct a lightning strike to ground. Better to "force" the thunderbolt's charge to hit the lightning rod, which was grounded, than to have it fall elsewhere on the building, in a place less protected. Wilson's reasoning on the rod's power of attraction indeed seemed at times questionable, as when he stated, after the committee decided that a pointed rod would operate at twelve times the striking distance from a cloud as a blunt rod, that a lightning stroke received on a pointed rod must therefore carry twelve times the quantity of electricity.

It was not unexpected when the Board of Ordnance ignored Wilson's views and, following the committee's advice, ordered lightning rods with points for the Royal Arsenal. The report reminded the Board that "Mr. Wilson's Objections to the Use of Pointed Conductors were heard . . . with Patience and Candour [and] were all answered and obviated to the Satisfaction of every one of the Committee but himself."

Wilson fought back by publishing a monograph, *Further Observations Upon Lightning,* in which he expanded on the notion that electrical charges held in clouds could be dangerously large, and that some would overpower any lightning rod. He took a poke at Franklin and the "Franklin rod," daring "The Philosopher" to walk or ride with a metal crown on his head "over Salisbury plain, in the midst of a very violent thunderstorm, notwithstanding there should drag behind him a communication of metal from the crown itself to the earth, during the whole of his journey." Wilson was, in effect, asking whether Franklin would be foolish enough to go about wearing on his own head the kind of lighting protection he recommended for important buildings.

Franklin, confident of his recommendations and a bit exasperated with Wilson's resistance, assured the Swiss physician and electrician Horace-Bénédict de Saussure:

> Pointed Conductors to secure Buildings from Lightning have now been in use near 20 years in America, and are there become so common, that Numbers of them appear on private Houses in every Street of the Principal Towns, besides those on Churches, Publick Buildings, Magazines of Powder, and Gentlemens Seats in the Country. Thunder Storms are much more frequent there than in Europe, and hitherto there has been no Instance of a House so guarded being damaged by Lightning.

There the matter rested for five years—until the night of May 15, 1777, when lightning struck and damaged the Board of Ordnance's Meeting House, on the grounds of the Purfleet arsenal but some distance from the gunpowder magazines. Wilson, claiming total vindication, blamed the pointed rods he had so vehemently opposed. King George III himself took new interest in the situation because, following the Board of Ordnance's lead at Purfleet, he had ordered Buckingham Palace outfitted with pointed rods. The Wilson-Franklin controversy thus stirred to life, and in a far different political climate from that in 1772.

Franklin had so enjoyed his stay in England from 1757 to 1762 that upon his return to Philadelphia he briefly—and perhaps half-heartedly—envisioned returning there to live, provided he could, among other obstacles, convince Deborah to set foot aboard a transatlantic ship, something she had always refused to do. His second mission there, from 1764 to 1775, was decidedly less pleasant. It was the time of the hated Stamp Act, increasing tension between the mother country and the colonies, and of growing doubt on Franklin's part that England would ever behave honorably, or intelligently, toward its "American plantations." The Stamp Act, which called for the colonists to pay a tax on all printed items, was seen in America as a serious escalation of London's impositions and was strongly resisted. "Upon the whole," Franklin wrote, "I have lived so great a Part of my Life in Britain, and have formed so many Friendships in it, that I love it and wish its Prosperity, and therefore wish to see that Union on which alone I think it can be secur'd and establish'd." But he warned that although many of the colonists felt similarly, "every Act of Oppression will sour their Tempers, lessen greatly if not annihilate the Profits of your Commerce with them, and hasten their final Revolt." While "kind Usage and Tenderness for their Privileges" would probably keep Americans loyal, he lamented that in London he did not "see here a sufficient Quantity of the Wisdom that is necessary to produce such a Conduct." Indeed, when the Stamp Act was finally repealed by Parliament, the British still demanded that the colonies pay for the stamps already printed, an injunction Americans found particularly contemptuous. (Franklin said it called to his mind the story of a man known to sit on the Pont Neuf in Paris brandishing a red-hot iron, asking passersby, "Would you kindly let me insert this up your arse?" When they angrily declined, he would say, "Well, all right, I won't insist on it, but at least pay me something for keeping the iron warm.")

The culmination of Franklin's estrangement from Britain arrived in 1774 when he was called before the English Privy Council to hear charges that he had improperly forwarded to a committee of patriots in Boston the private correspondence of Thomas Hutchinson and Andrew Oliver, Americans who were, respectively, the governor and lieutenant governor of Massachusetts. Among the provocative information in the letters was the view that Britain would likely need to dispatch troops to North America. Although Franklin's intention may have simply been to show his own countrymen that differing opinions about affairs in the colonies were held by men on both sides, the letters' publication in Boston led British authorities to accuse him of acting deceitfully in order to stir up bad blood between the Crown and the provinces. Publicly denounced in a humiliating dressing-down by Alexander Wedderburn, England's solicitor general (who, in a jab at Franklinian science, called him the "prime conductor" of the Hutchinson letters affair), Franklin was stripped of his office of postmaster general of the colonies. His reputation tarnished, and his usefulness in London now greatly weakened, he sailed home in March 1775, remarking to friends who saw him off at Portsmouth, "It seems that I am too much of an American."

While he was at sea, on April 19, 1775, the first shots of the Revolutionary War were fired at Lexington and Concord. The following July the Declaration of Independence was read from a balcony of the State House in Philadelphia. By Christmas 1776, with Britain and the colonies at war, Franklin was in Paris preparing to represent the embryonic American government to the court of France.

Although Wilson had won a Copley Medal in 1760, the award had come as a result of his exacting experimental work; as a theorist he was regarded as "a bit muddled" by his scientific peers. No doubt aware of the condescension, Wilson developed a marked distrust of his colleagues, having once accused Watson, a former mentor, of plagiarizing his work. But Wilson was well-connected, his prolific portraiture business having allowed him to befriend members of the royal family. With Franklin no longer in England, and the king newly sympathetic to Wilson's claims, the monarch gave his approval when Wilson suggested an elaborate means of demonstrating his argument for blunt rods—the building of a giant "artificial cloud" in the London Pantheon, a dance pavilion often used for masquerade balls.

Wilson's "lightning simulation device" consisted of an electrified "cloud," a cylinder 16 inches in diameter and 155 feet in length, suspended by silk and poised over a scale model of the Board of Ordnance's Meeting House at Purfleet, which was armed with both miniature pointed and blunt rods and situated on a trolley, the movement of which approximated a storm's approach. In repeated experiments, Wilson bombarded the model with shocks and observed that the pointed rods drew the electrical discharge from the cylinder at a greater distance. In Wilson's view, this indicated the more substantial danger they posed. Franklin's loyalists, led by Charles, Viscount Mahon, the third Earl of Stanhope, countered that this power to draw the electrical fire from the clouds while they were still some distance away diminished the potential strength of any lightning bolt that might strike the arsenal, and made it more likely any and all would be intercepted, thus offering better protection.

The London newspapers closely followed the Pantheon experiments, which were open to the public and the press. It was noted disapprovingly that Wilson often repaired to the far end of the room when the Franklinists made their counter-demonstrations, and Wilson's findings were ultimately labeled "inconclusive" by the Royal Society. Privately, among his scientific brethren, there was harsher criticism—that his demonstrations were deceptive and his conclusions fraudulent. Franklin's friend Jean-Hyacinthe de Magellan, an expert on scientific instruments, minced no words, castigating the "Mischievous Wilson" as a "puffing . . . pretended Philosopher."

Given the political sentiments of the time, however, and the fact that the Purfleet facility had suffered damage with a pointed "Franklin rod," the king declared himself persuaded by Wilson's proofs. "They are so plain," he remarked, "they would convince the apple-women of Covent Garden." He ordered pointed rods removed and replaced with blunt ones atop Buckingham Palace and the other royal residences. The king could not literally dictate the choice of rods on the arsenal—this was Parliament's responsibility—but his strong public statements on the affair did hinder the further deployment of pointed lightning rods at Purfleet. Tragically, they had the same discouraging influence at a British magazine in the colonial outpost of Sumatra which, along with its four hundred barrels of gunpowder, was soon after blown up by lightning.

Franklin, learning of the royal decree Wilson had won, seems only to have been amused. "I have no private interest in the reception of my inventions by the world," he wrote, "having never made, nor proposed to make, the least profit by any of them. The King's changing his pointed conductors for blunt ones is, therefore, a matter of small importance to me. If I had a wish about it, it would be that he had rejected them altogether as ineffectual. For it is only since he thought himself and his family safe from the thunder of Heaven, that he dared to use his own thunder in destroying his innocent subjects." The king, perhaps at Wilson's urging, then appealed to the Royal Society to withdraw its earlier affirmation of Franklin's electrical work. This request the Society tactfully dodged, its leader, Sir John Pringle, being heard to comment, "The prerogatives of the President of the Royal Society do not extend to altering the laws of nature." The king's ruling was also mocked in a popular ditty:

> While you, George, for knowledge hunt,
> And sharp conductors change for blunt,
> The Nation's out of joint:
> Franklin a wiser course pursues,
> And all your thunder useless views,
> By keeping to the point.

Remote as the passions of the Wilson-Franklin dispute seem now, the debate about pointed versus blunt-ended lightning rods has proved among the liveliest and most durable in the physical sciences. While glow-discharges are easier to create with sharp points than blunt in a laboratory setting, against the vast natural scale of an actual thunderstorm the effectual difference between pointed and blunt rods is negligible, although lightning safety councils recommend, and most manufacturers continue to make,

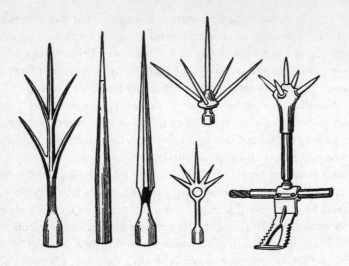

pointed lightning rods. (Physicist Philip Krider, one of America's top lightning experts, notes that *all* lightning rods, once struck by lightning, become blunt-ended.) The very question of whether lightning rods neutralize passing clouds, drawing the "fire" from them, as Franklin once believed, is considered fairly moot by present-day physicists. Experience has shown that only a rod situated at a very great height, higher than most installed lightning rods, will have this effect. By and large the value of lightning rods, pointed or blunt, is in their ability to serve, if properly installed and grounded, as an attractive receiving terminal for a thunderbolt that would otherwise have struck a protected structure, safely conducting it into the earth.

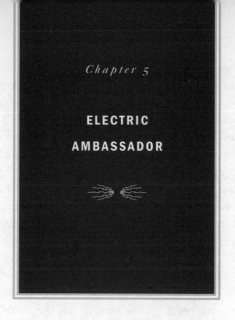

Chapter 5

ELECTRIC
AMBASSADOR

BECAUSE OF HIS SCIENTIFIC FAME ABROAD AND HIS LONG EXPERIENCE as a colonial agent in London, Franklin was a natural choice to help represent the American Revolution's ideals to sympathetic figures in Europe. In May 1776, France, seeing in the colonists' resistance to Britain a chance to harass their longtime enemies while improving their own stature in the New World, offered to contribute one million livres to the rebels' cause. The aim of Franklin, who arrived in Paris at Christmas 1776, and of his fellow commissioners representing Congress to the Court of France, Silas Deane and Arthur Lee, was to facilitate and hopefully enlarge on this act of generosity. Seventy years old in the year the colonies declared themselves a nation, Franklin assumed his new role with self-effacement and aplomb, telling his friend Benjamin Rush, "I am like the remnant of a piece of unsalable cloth you may have, as the shopkeepers say, for what you please."

There was a charming can-do-ism about the United States' first efforts at national diplomacy. Even Franklin, who, for an American of his day, was extremely well-traveled, had little experience as an actor in European courts. However, in "address and good breeding," recorded John Adams, who joined the American delegation in Paris in 1778, when Franklin was made minister plenipotentiary to France, "he was excelled by very few Americans." And Franklin's mind, at seventy, was still a fine and supple tool, his thinking honed by years of civic leadership, negotiation, and exacting science.

Franklin had not originally advocated separation from the mother country, but once the choice was made, he became an ardent believer in both independence and the future of an American republic. Bolstering his faith in the rightness of the American cause was his observation that the colonies would soon have a majority of the English-speaking peoples of the world, and would surely no longer need to cede authority over their affairs to a nation three thousand miles away. Still, with colonial manufacturing long restricted by British edict and, of course, no hope of obtaining supplies from England, the material needs of the colonists at the war's beginning were enormous. Franklin emphasized to his countrymen the want of ammunition, artillery, muskets, and "good engineers" to build breastworks and ports.

With not enough guns to go around, he helped design a pike for military use. In a memorandum to the Pennsylvania Assembly he explained the weapon's simple and deadly battlefield effectiveness throughout centuries of human conflict, and advised, "Every Smith can make these, and therefore the Country may soon be supply'd with Plenty of them." He also advocated the use of bows and arrows, because "a Flight of Arrows seen coming upon them terrifies, and disturbs the Enemy's Attention to his Business." Since arrows had proved such an effective weapon in the Middle Ages, when combatants wore body armor, Franklin asked, wouldn't they be even more deadly today? Of course Franklin knew well the answer to his own question: pikes and arrows could not be counted even as half-measures in a war against the cannons and infantry of one of the world's great military powers. He was simply trying to be encouraging. Yankee ingenuity alone, he knew, would not deter the king's armies.

Franklin arrived in Paris at an auspicious moment, for the French had recently developed a keen fascination with things American. Books and articles proliferated about the infant nation's natural beauty, the industriousness and fair-mindedness of its people, and the habits of a free and magnanimous society now in defiant revolt against tyranny. In the strivings of the young American republic, the Old World saw "the project of enlightenment in practice"—a program it knew for the moment to be beyond its own grasp, but one that was inspiring to watch nonetheless. Jacques Turgot declared, "America is the hope of the human race, and can become its model," a sentiment with which many Frenchmen would have concurred, although Turgot himself, serving as France's controller-general, had been one of the few ministers to Louis XVI who cautioned against involvement in the Americans' war. Franklin, meanwhile, proved to be a delightful enigma to his hosts. Eschewing the powdered wig fancied by the genteel classes, he went about in a fur hat and "Quaker" shoes that accentuated his rusticity, a good wizard from North America who feared neither lightning nor kings. Franklin's fame as the inventor of the lightning rod became as one with his role as a member of the Continental Congress and a signer of the Declaration of Independence—an association the late eighteenth century found irresistible and that became immortalized in Turgot's epigram, "*Erepuit caelo fulmen sceptrumque tyrannis*" ("He snatched lightning from the sky and the scepter from tyrants").

Turgot's interest in Franklin's electrical discoveries went beyond favorable phrases. He and his fellow French economists—the Physiocrats—saw in Franklin's theories of electricity an analogy supportive of their own theories of economics, for they believed that wealth, like electricity or other subtle fluids, was capable of promoting equilibrium in society. This "natural" economic law needed only noninterference to succeed, as connoted in the Physiocrat motto, *laissez-faire* ("let it be"). Perhaps this was the appeal of Franklin's electrical ideas for progressive minds: They described not a cold, mechanical process but one that seemed *alive*, one in which nature seemed to act with purpose and restraint, as though out of some innate sensibility, to achieve balance. With France's ugly and seemingly irreparable social disparity between

the nobility and the wretched, the Physiocrats and other would-be reformers must have been captivated by Franklin's description of America as a place of "happy mediocrity," one "ruled" by "the middling people," where no one was extravagantly rich or desperately poor, but all worked for their living. That Franklin himself was of common stock, and self-educated, was of course central to his legend. "Now one of the first characters in the philosophical and political world," the French noted, "[he] owes his present elevated rank in life entirely to himself."

Nor was this latter fact ever lost on Franklin, who took great pride in how far he had come from his "low beginning." A biblical proverb his father had often repeated to him as a child, "Seest thou a Man diligent in his Calling, he shall stand before Kings, he shall not stand before mean Men," resonated deeply with Franklin the adult. When in 1768 he dined in England with King Christian VII of Denmark, he was so thrilled by the occasion that he wrote a detailed account and even drew a diagram of the seating arrangements. He cherished the honorary doctorate St. Andrew's University had bestowed upon him in 1759, and took great pride in being addressed as "Doctor Franklin." In France he soon had the added distinction of becoming a bestselling author. A selection from *Poor Richard's Almanac* known as "The Way to Wealth" (or "Father Abraham's Speech"), an engaging compendium of Franklinian wit counseling thrift, industry, and virtue, produced in 1757 in honor of *Poor Richard's* twenty-fifth anniversary, was translated as "La Science du Bonhomme Richard" and sold briskly, turning many of his maxims into household expressions. "He that lives upon Hope will die fasting," Father Abraham, the fictitious wise man, reminds Franklin's readers. "The Used key is always Bright . . . the Cat in gloves Catches no Mice . . . God helps them that help themselves . . . In the Affairs of this World, Men are saved, not by Faith, but by the Want of it."

France duly treated him as a celebrity. The man who had understood the importance of his fellow Philadelphia tradesmen seeing him push his own paper in a wheelbarrow perceived also the value of playing the role of the Yankee rustic for the French, and he would not have been displeased to know he'd been described by a courtier as having visited Versailles "in the dress of an American farmer." His face and profile began to appear on medallions and the lids of snuffboxes, and the famous Jean-Antoine Houdon executed a bust of him. Franklin's likeness often contained a fancified image of the kite experiment or of his using lightning bolts to destroy the symbols of monarchy. In honor of his great invention, *le chapeau paratonnerre,* or "the lightning rod hat," appeared for ladies, complete with trailing chains for "grounding," as did later the *parapluie-paratonnerre,* a gentleman's umbrella.

Since Franklin had shown that thunder and lightning were natural processes, composers began to allude musically to "thunder" in arrangements for their operas, and exhibitors of magic-lantern shows installed it as a sound effect, along with other audible imitations of nature. As the lightning bolt had once been God's weapon of displeasure, it now became the people's, and appeared frequently in broadsides and pamphlets depicting the power of liberation against the polluted institutions of church and government.

Thunderbolts were shown knocking kings and princes off their thrones and, prophetically, blasting to bits the hated Bastille. Lightning might continue to occasionally exact nature's revenge, but it would never again hurt the people. "If Jupiter wants to blast us to bits, wise Franklin," read one assured caption, "you remind him of your laws."

"Franklin's own popularity was so widespread that it does not seem exaggerated to call it a mania," historian Simon Schama has suggested. In June 1779 he wrote his daughter, Sally, that the countless likenesses of him "have made your father's face as well known as that of the moon." And his renown was no passing fancy. When Thomas Jefferson got to Paris in 1785, nine years after Franklin's arrival, he saw that "there appeared to be more respect and veneration attached to the character of Dr. Franklin in France than to that of any other person in the same country, foreign or native." Perhaps the only skeptic was Louis XVI, who it was said became so tired of listening to a mistress recite Franklin's praises that he presented her with a chamberpot with a medallion bearing Franklin's image at the bottom.

Franklin likely thought the French slightly mad in their adulation. He became used to saying, in response to hearing Turgot's oft-repeated epigram, that he had left thunder where he'd found it, and that more than a million of his countrymen had helped him snatch a king's scepter—but he was enough of a natural diplomat to understand the value of allowing a people whose help his country desperately needed to project onto him whatever positive attributes they wished. "The Franklin legend answered the intellectual and sentimental needs of the French public," André Maurois wrote. "There was nothing rustic about Franklin, and he was far more sharp than simple; but he was beautifully able to play the part which was assigned to him."

The apotheosis of Franklin as the democratic messiah occurred in April 1778, when he joined Voltaire at a meeting of the Academy of Science. As the two great living symbols of the Enlightenment bowed to one another, the cheers of those present grew deafening. In response they clasped hands. Still the onlookers clamored, until someone made their demand clear: "*Il faut s'embrasser, à la française!*" Recalled John Adams, who was present, "The two Aged Actors upon this great Theatre of Philosophy and frivolity then embraced each other by hugging one another in their Arms and kissing each other's cheeks, and then the tumult subsided." Voltaire told Franklin that he held what was taking place in America in such high regard that, had it occurred forty years earlier, he would have been eager to "establish himself in so free a country."

Adams, relegated to observer status as his famous colleague was everywhere hailed and cheered, later would remark with a mixture of envy and admiration that "It is universally believed in France that his electric wand has accomplished all this revolution." He feared history would recount that "Dr. Franklin's electrical rod smote the earth and out sprung General Washington. That Franklin electrized him with his rod—and henceforward these two conducted all the policy, negotiation, legislation and war." Adams esteemed Franklin but never much liked or trusted him, or perhaps

he simply was annoyed by the ease with which Franklin seemed to float through life, effortlessly garnering laurels and approbation. "I can only suggest that Adams must have felt toward him the way Salieri did toward Mozart," notes Franklin biographer Claude-Anne Lopez.

The tension between the flinty Adams and the genial Franklin dated at least from the fall of 1776, when they were forced to share a bed at an inn in New Jersey en route to a parlay with the British. The fresh air–loving Franklin insisted on leaving the window open all night, while Adams thought it best that it be closed. Franklin, a believer in the healthy regimen of a daily "air bath" ("I rise almost every morning and sit in my chamber without any clothes whatsoever, half an hour or an hour, according to the season, reading or writing . . ."), felt certain that "people often catch cold from one another when shut up together in close rooms, coaches, &c." According to Adams, Franklin got his way and the window was left open, although "the Doctor then began an harangue" about the benefits of fresh air that grew so tedious it put Adams to sleep. (Adams, not one to leave an argument unfinished even after a number of years, asserted [incorrectly] after Franklin's death that the great man had died "a Sacrifice . . . to his own theory; having caught the violent cold, which finally choked him, by sitting hours at a Window, with the cool Air blowing upon him.")

What seemed to irritate Adams upon his arrival in France was the extent to which Franklin appeared to have "gone native," taking up not only French habits but also the country's loose mores, and becoming devoted to a life of indulgence—late suppers, soirees, all-night chess games, and the company of some imposing French women as confidantes and intimate friends. Adams, by his own admittance, had nothing but contempt for "balls, Assemblies, concerts, cards, horses, dogs, which never engaged any part of my attention." He was equally unimpressed by the *philosophes* Franklin admired, finding their dreams of human improvement naïve and as ineffectual as any religious enthusiasm.

"Somebody, it seems, gave it out that I lov'd Ladies," Franklin wrote to a friend from Paris, "and then every body presented me their Ladies (or the Ladies presented themselves) to be embrac'd,

that is to have their Necks kiss'd. For as to kissing of Lips or
Cheeks it is not the Mode here; the first is reckon'd rude, & the
other may rub off the Paint. The French Ladies have however
1000 other ways of rendering themselves agreeable; by their various Attentions and Civilities, & their sensible Conversation. 'Tis a
delightful People to live with."

Lopez uses the term *amitié amoureuse* to characterize Franklin's
friendships with women, a form of mutual devotion distinguished
by flirtation, affectionate correspondence, and occasional light
snuggling that generated more warmth than heat. Adams, however, may have seen (or suspected) more. "The temple of human
nature has two great apartments: the intellectual and the moral,"
Adams wrote. "If there is not a mutual friendship and strict alliance
between these, degradation to the whole building must be the
consequence."

There likely was more validity in another of Adams's concerns—that Franklin appeared overly nonchalant about the delegation's mission. Separated by the Atlantic Ocean from an
American war effort that often seemed hopeless, Franklin chose to
handle the formidable task of diplomacy by remaining pointedly

low-key, more like a well-mannered guest than a skilled envoy. As ever, the key to Franklin's success was his effortless adaptability, his knack for remaining just above the fray of ordinary human affairs even as, with his words and actions, he demonstrated a firm understanding of human nature. "If Passion drives, let Reason hold the Reins," *Poor Richard's* had cautioned. Franklin understood that self-control, moderation, and compromise were not only the means to his own personal happiness, but tools the rest of the world might use to its own great benefit. He preached their advantages in his almanacs, pamphlets, and public appeals, and even in the gentle hoaxing through which he sought novel ways to impart truth. Franklin could be very clever, but he was also wise, and he seldom lost sight of the distinction between cleverness and wisdom. In France, his determination to evince a steady optimism, no matter how dire the news from home, was perhaps best illustrated by his response to the courier who awakened him early one morning with the news that the British general William Howe had taken Philadelphia. Franklin, it was said, raised himself on one elbow, then corrected his visitor, "No, Monsieur, it is Philadelphia that has taken Howe!"

Not that Adams's more aggressive efforts at diplomacy fared better. When he attempted to approach the Comte de Vergennes, France's foreign minister, in a more forceful style than had Franklin, Vergennes complained to Franklin that Adams was a nuisance and for a time refused to meet with him. "[Mr. Adams] is always an honest Man, often a wise one," Franklin concluded, "but sometimes . . . absolutely out of his senses."

Franklin was less tolerant of another fellow commissioner, Arthur Lee, who complained vociferously to Franklin (and to Congress) that Franklin was not fully sharing information about America's diplomatic strategy. His reply to one of Lee's accusatory notes offers something rarely glimpsed—Franklin at full pique:

> It is true I have omitted answering some of your Letters. I do not like to answer angry Letters. I hate Disputes. I am old, cannot have long to live, have much to do and no time for Altercation. If I have often receiv'd and borne your

Magisterial Snubbings and Rebukes without Reply, ascribe it to the right Causes, my Concern for the Honour and Success of our Mission, which would be hurt by our Quarrelling, my Love of Peace, my Respect for your good Qualities, and my Pity of your Sick Mind, which is forever Tormenting itself, with its Jealousies, Suspicions and Fancies that others mean you ill, wrong you, or fail in Respect for you. If you do not cure your self of this Temper it will end in Insanity, of which it is the Symptomatick Forerunner, as I have seen in several Instances. God preserve you from so terrible an Evil, and for his sake pray suffer me to live in quiet.

Franklin set the letter aside and, thinking he had been too harsh, tried again the next day, with slightly milder results. In all he wrote three drafts, none of which he ever sent.

In his ability to acclimate to new surroundings, Franklin was a bit like the polyp he so admired: Cut him off from his home and family in Philadelphia and place him in London, he regenerated as an English gentleman; plunk him down in Paris and he became, with his wit and earthy charm, an admired Yankee *philosophe,* and one with a substantial wine cellar, no less. Regardless of where he was, Franklin was always by nature an inventor, someone whose rich imagination allowed him to perceive a chain of causation and then visualize potential solutions more readily than others.

As his power of sight diminished with age, he had grown frustrated that he had to be constantly switching between two pairs of spectacles. When he glanced up from his work, objects in the near-distance refused to come into focus, a difficulty that became most acute when he was riding in a carriage. "I sometimes read, and [yet] often wanted to regard the Prospects," he said. Of even greater concern, because it affected his diplomatic mission, was his discovery that at the dinner parties he so enjoyed, one pair of glasses would not suffice to let him see both the food on his plate, and the eyes,

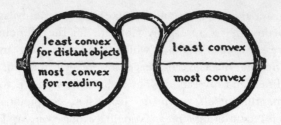

lips, and gestures of his hosts, a practice essential to his under-standing what was being said, since his French was adequate at best. Returning home in frustration from one such experience, he hit upon the solution: He would cut two separate pairs of his own glasses into semicircles, and then combine their lenses. "By this means," he wrote to a friend, George Whatley, "as I wear my spec-tacles constantly, I have only to move my Eyes up or down as I want to see distinctly far or near, the proper glasses being always ready." Proud of how, with a bit of ingenuity, he had conquered at least one imposition of old age, Franklin mused that "if all the other Defects and Infirmities were as easily and cheaply remedied, it would be worth while for Friends to live a good deal longer."

At around the same time Franklin was inventing bifocals, he (and everyone else in Paris) became enthralled by a far more dra-matic innovation—balloon flight.

The first balloon ascension, at Annonay on June 4, 1783, was staged by the brothers Joseph-Michel and Jacques-Étienne Montgolfier. Papermakers by trade, they reportedly had the inspi-ration for their linen-and-paper balloon after one of them observed a piece of clothing lifted into the air by the heat from a fireplace. This would have conformed with an idea then widely accepted that clouds were pockets of air warmed by the sun and, thus rar-efied, rose into the sky. What would result, the Montgolfiers asked, if such a pocket of heated air was to be captured in a silk balloon?

Franklin was a swift convert to ballooning, and an ardent en-thusiast. On August 27, he was present along with an estimated fifty thousand people at an unmanned ascension staged by Jacques Charles on the Champ de Mars. Instead of filling his balloon with hot air as the Montgolfiers had, Charles used hydrogen gas, discovered in 1766 by the English scientist Henry Cavendish, which was created by putting sulfuric acid onto iron filings. Hydrogen balloons were called *Charlieres*, in contrast to the hot air–powered *Montgolfieres*. The crowd, along with Franklin, gazed in wonder as the tiny *Charliere*, only twelve feet in diameter, rose to a height of about fifteen hundred feet and glided across the Seine. "It diminish'd in Apparent Magnitude as it rose," Franklin reported in a letter to Joseph Banks of the Royal Society, "till it enter'd the Clouds, when it seemed to me scarce bigger than an Orange, and soon after became invisible, the Clouds concealing it." The balloon sailed twenty kilometers north of Paris and returned to earth after a forty-five-minute flight. The ascension was the setting for one of Franklin's most treasured quips. Overhearing a skeptic in the crowd question what actual use balloons might have, Franklin replied, "What is the *use* of a new-born baby?"

On September 19 of that year, the Montgolfiers staged a more daring ascension at Versailles before a crowd of 130,000, including the entire royal family and the retinue of the court. The balloon, carrying the world's first air passengers—a sheep, a duck, and a rooster—rose to a height of fourteen hundred feet, traveled about four kilometers, and fell to earth after a flight of ten minutes. Veterinarians on the scene quickly examined the balloon's animal cargo and declared all to be in good health.

The first manned flight came on November 21 in a *Montgolfiere*, when two "courageous philosophers"—Jean-François Pilâtre de Rozier and the Marquis François d'Arlandes—went up from the garden of La Muette, the queen's palace in Passy. De Rozier and d'Arlandes waved their hats at the spectators, including Franklin, who lived nearby, as they flew to fifteen hundred feet. "There was a vast Concourse of Gentry in the Garden," Franklin wrote, "who had great Pleasure in seeing the Adventurers go off so cheerfully, and applauded them by clapping, &c. but there was at the same time a good deal of Anxiety for their Safety." On a stiff westerly wind the balloonists traveled for almost twenty minutes to a gentle landing outside Paris on the Fontainebleau road, where they served glasses of champagne to the anxious farmers who gathered around the fallen craft. De Rozier and his companion were fortunate: The

Charles balloon of August 27 had fallen unannounced in the village of Gonesse, twelve miles from Paris, and frightened peasants "conceiv'd from its bounding a little, when it touched the Ground, that there was some living Animal in it, attack'd it with Stones and Knives." In response, Louis XVI issued a proclamation against assaulting balloons.

On December 1, Franklin, the king, Marie Antoinette, and other notables were present to watch Charles in another manned ascent. A throng of four hundred thousand, then considered the largest gathering of humanity ever assembled, filled the Tuileries and the banks of the Seine. In the early evening light, Charles and two companions rose to an unprecedented height of nine thousand feet and remained aloft for almost a half hour. "Never before was a philosophical experiment so magnificently attended," Franklin commented, writing happily of the flight and noting that the wicker car beneath the balloon was large enough for a small table at which the balloonists could sit and "keep their Journal." Upon landing, Charles's own euphoria was so great that he insisted on going up again immediately by himself, becoming, as one historian has pointed out, "the first man to see the sun set twice in one day."

To Banks, Franklin cited the lively competition between the two kinds of balloons and their brave inventors, and the rapid developments that had resulted. "And one cannot say how far it may go," he concluded. "A few Months [ago] the Idea of Witches riding thro' the Air upon a Broomstick, and that of Philosophers upon a Bag of Smoke, would have appear'd equally impossible and ridiculous."

Flight, so long a fantasy of man, caught France by storm, with a frenzy probably not unlike that with which Americans responded to the appearance of the first airplanes in the early twentieth century, an ardor akin to religious enthusiasm. The maneuvers of the colorful fabric balloons over Paris made starry-eyed children of kings and natural philosophers alike, while fashionable women donned *chapeaux au ballon* and journalists penned biographical tributes to the lives of the daring balloonists and wrote ecstatically of the otherworldly sensation of flying. Particularly fascinating to readers were the balloonists' descriptions of how the earth appeared from the sky, and what the tops of clouds looked like.

Franklin was as proud as any Frenchman of the balloonists' accomplishments, even chiding Joseph Banks for allowing France to jump out to such an early lead in so inviting a field. "Your Philosophy seems to be too bashful," Franklin wrote. "In this Country we are not so much afraid of being laught at. If we do a foolish Thing, we are the first to laugh at it ourselves, and are almost as much pleased with a Bon Mot or a good Chanson, that ridicules well the Disappointment of a Project, as we might have been with its Success." (Here Franklin seems to be describing not so much the French as himself, although in his mind the two had apparently become one.) Drawing an imaginative analogy, he reflected: "We should not suffer Pride to prevent our progress in Science. Beings of a Rank and Nature far Superior to ours have not disdained to amuse themselves with making and launching Balloons, otherwise we should never have enjoyed the Light of those glorious Objects that rule our Day and Night, nor have had the Pleasure of riding round the Sun ourselves upon the Balloon we now inhabit."

In an exchange of letters with a Dutch friend, physician Jan Ingenhousz, Franklin discussed the hope that balloons would create "perpetual peace," since in a war, an attack from the air by hundreds of men in balloons would render all earthbound defenses useless. Faced with five thousand balloons, each bearing a crew of two men, "where is the prince who can afford to cover his country with troops for its defense as that ten thousand men descending from the clouds might not in many places do an infinite deal of mischief before a force could be brought together to repel them," Franklin asked. Ingenhousz agreed, terming balloons "a discovery big with the most important consequences and capable of giving a new turn to human society." Of London's silence on the success of the French ascents, Ingenhousz confided, "I believe it is national jealousies which make the English look upon the air balloons with so much indifference."

The excitement soon caught on in England, however. On January 7, 1785, John Jeffries, a native of Boston now residing in London, and Jean-Pierre-François Blanchard, who would later invent the parachute, crossed the English Channel from Dover to Calais in a *Charliere*. Partway across, their balloon began to lose altitude. With the frigid gray waters of the Channel looming only feet below, the two men began to desperately hurl items out of the balloon, including some of their own clothes, in an attempt to regain altitude. Their quick action no doubt saved their lives. After managing to land in France, Jeffries turned up in Passy bearing a piece of correspondence from England—the world's first "airmail letter." The delighted recipient, Dr. Benjamin Franklin, insisted the postman stay for dinner.

In France, of all places, one might have expected the lightning rod to gain immediate acceptance. Buffon, the "French Pliny," had endorsed it; D'Alibard, with the help of the alert dragoon Coiffier, had provided the first demonstration that clouds were electrified; and the famous Franklin, "*l'ambassadeur électrique*," who challenged thunderbolts and kings, had become a fixture of Parisian life. As had the British before them, the French enlisted Franklin in 1784 to advise the government how best to protect powder magazines. An effort he joined to consider the safety of the magazine at Marseilles produced science's first consensus about the physical area safeguarded by an individual lightning rod. The English, when they used lightning rods, tended to festoon rooftops with them, believing this guaranteed the best coverage; the French, seeking greater precision, determined that an individual lightning rod protected an area within a forty-five-foot radius.

In France, however, as in Boston and in London, qualms about the *paratonnerre* arose despite the continuing acclaim enjoyed by its inventor. There were in fact few lightning rods on people's private dwellings in France, as buying and mounting the devices could be expensive; they tended to be used chiefly by churches, at government or military installations, or by the owners of prominent châteaux. However, unlike in England, where the question of blunt or pointed rods had set off an ugly squabble involving Buckingham Palace and the Royal Society, or in New England, where the lightning rod had raised concerns about presumption before God, in France the debate played out over the rights of one solitary, stubborn property owner. The resulting legal case would give the lightning rod its only real "day in court," and bring to prominence a young man whose fate would be intimately bound up with his country's, Maximilien de Robespierre.

In June 1780, M. Vissery de Bois-Valé, a resident of Saint-Omer, a village between Calais and Lille, was approached by concerned neighbors who insisted he remove a lightning rod he had placed atop his roof. When he refused, they won an injunction from a local magistrate, who ordered him to take the rod down within twenty-four hours. Vissery was a retired lawyer and a small-time inventor who had devised a method of keeping fresh water potable, and a primitive apparatus to enable divers to breathe underwater. Well-off, he could afford to dabble in experiments and stay abreast of important trends in natural philosophy; he was proud to have Franklin's technology protecting his property. His lightning rod was highly original, consisting of a large sword mounted on a weathervane whose base was an artistically rendered globe being struck by lightning and shooting forth jagged rays. The globe itself was mounted on a sixteen-foot iron bar, at the bottom of which a fifty-seven-foot-long "tail" led across Vissery's rooftop and down his and an adjoining neighbor's wall to a rod and chain device that disappeared into a well. Local aldermen who investigated described the arrangement as one in which "the thunder will be allowed to descend . . . and drown itself." But his poorer neighbors resented his arrogance in presuming to put them at risk.

This was not the first time Europe had seen citizens turn on a "Franklinist" in their midst. In Primetice, Moravia, in June 1754, a monk and amateur scientist named Procopius Divis had put up his version of a lightning rod, a "meteorological machine" that consisted of a pole topped by an iron rod upon which rode twelve "branches," each with a small box filled with iron ore. The Divis apparatus was probably more sculpture than functional lightning rod, although in 1760 the villagers, uninterested in such distinctions, ripped it down, blaming it for causing a recent punishing drought. Even in the enlightened university town of Bologna, Italy, electrical experimenter Giovan Veratti was hounded into removing the rods he'd established on some school buildings, despite a letter from Pope Benedict XIV recommending their use.

In the Saint-Omer case, the problem started with a disagreeable neighbor on the rue Marche-aux-herbes, Mme. Renard-Debussy, whom Vissery regarded as "an old quibbler" and with whom he had never enjoyed favorable relations, and her friend Mme. Cafieri. The women may have had their fears of lightning exacerbated by an uncanny occurrence in nearby Arras a few years earlier, when a church steeple was struck with such force by a thunderbolt that the flagstones of the church portico were jarred loose and observed to fly through the air. Debussy and Cafieri, according to Vissery, had led a campaign to rouse other women (and a few men) to their cause, and had delivered a petition "dictated by ignorance" to the local magistrates. Vissery decried the magistrates' injunction against his apparatus as having been issued without regard "for the cited authority of the greatest *physiciens,* Academies, republics, and entire realms that have adopted this admirable invention." Instead, the bureaucrats had mindlessly joined the petitioners in showing "hostility toward the *monde savant.*" At one point, when townspeople gathered outside his house to gawk at the contraption on his roof, Vissery angrily rushed out to confront them with an edition of the *Journal de Physique* that described the usefulness of lightning rods.

Two days after the injunction was handed down, Vissery announced his intention to resist its order, and on June 21 arguments were made before the local aldermen. They, unlike some of the villagers, did not find a lightning rod to be a presumption against God, but did question Vissery's ability to mount such a device properly. There existed at the time a substantial, and mostly apocryphal, body of malfunctioning-lightning-rod stories—of rods melting or turning red hot, sparks shooting from their tips—and the aldermen voiced concern that because the rod was near Vissery's chimney, a lightning bolt that was attracted to the rod might engage the fire burning in the hearth, leading to unimaginable devastation. They reaffirmed the order that Vissery would have to take the rod down.

Vissery wrote of their "barbaric despotism" to Antoine-Joseph Buissart, a lawyer and regular contributor to journals of natural philosophy, who had once called lightning rods "the happiest discovery made this century." Buissart had become enamored of lightning rods while advocating their use to protect crops and vineyards from costly damage due to hailstorms. Countless attempts to prevent the frequent ravages of hail had been made for generations in rural France, from firing cannons into the air to the somewhat less scientific practice of walking the boundaries of one's property carrying a tortoise upside-down. The lightning rod, Buissart believed, offered farmers their first rational hope in the battle against ruinous hail. Vissery informed Buissart that he had taken down the rod as ordered, although he boasted in an aside that he had only made it appear that he had done so by replacing the large sword with a smaller one, confiding, "This is how to deal with the ignorant masses."

Buissart agreed to take Vissery's case, and dashed off letters seeking advice and support from several experts, including Franklin's friend Jean Baptiste Le Roy, the physicist, and the Marquis de Condorcet, a renowned mathematician and *philosophe*. Le Roy expressed to Buissart his incredulity that "fifty leagues from the capital and at the end of the 18th century," a dispute like that occurring at Saint-Omer was even possible. Condorcet advised Buissart that a thorough scientific report on Franklin's lightning rod would be useful in court, prompting Buissart to compile a list of all the lightning rods in use across Europe. The abbé Pierre Bertholon de St. Lazare, a designer of lightning rods who had proposed they be erected not solely to protect buildings but also to deter earthquakes and volcanoes, provided Buissart with an inventory of the eleven lightning rods he knew to be in use in France: one in Valence, three near Dijon, one in Bourg-en-Bresse, one in Anjou, four in Lyon, and one on Voltaire's house at Ferney. ("There are some great lords which it does not do to approach too closely," Voltaire had said, "and lightning is one of these.") Vissery, in the meantime, obtained from Louis-Bernard Guyton de Morveau, author of the article on "Thunder" in the 1777 edition of the *Encyclopédie*, a written certification that his lightning rod had been constructed in an approved manner. Other *philosophes*, writing to Buissart from Paris, urged him to remind the local authorities that they would be the butt of endless jokes and ridicule if they were to uphold a decision that a man could not put a lightning rod on his own rooftop.

Buissart eventually accumulated enough information to create a sixty-eight-page brief whose main thrust was that in cases such as Vissery's, in which legal and scientific authority were opposed, the former must give way to the latter. After all, the aldermen had not proven that lightning rods didn't work, or that Vissery's was improperly built. Buissart also suggested that the anxiety about lightning rods stemmed from the infamous death of Richmann in St. Petersburg in 1753, and that such fears were misplaced, since Richmann had been killed while using a "gnomon," a device unlike a grounded lightning rod, to measure the electrification of the atmosphere. Urging the authorities to allow "savants" to determine the fate of Vissery's rooftop appliance, Buissart compared it to a broken chimney: If a man said his neighbor's chimney was rickety, a judge wouldn't order it torn down without seeking a bricklayer's opinion.

Of course, Buissart was fudging a bit. The actual operation of a lightning rod was, even as late as the 1780s, only partly understood. Did the rod quietly diffuse electrified clouds passing overhead, its point drawing off sparks as in Franklin's experiments with a bodkin, or did it serve only to safely conduct actual bolts of lightning to the ground? Did it attract lightning to structures that otherwise would have been harmlessly passed over? Were sharp points preferable to blunt? Was the forty-five-foot radius of protection established by the Marseilles committee accurate, or was the zone actually smaller, or larger? In short, Buissart's argument that the matter should be left in the hands of science was somewhat strained, since the science itself was not entirely confident. Franklin, though in France at the time, did not apparently weigh in on the case, probably because he was immersed in diplomatic activities, but perhaps because it involved issues he believed to have been thoroughly adjudicated and that by now should "shift for themselves." He had never lost his "extreme Aversion to Public Altercation on Philosophic Points," preferring that his "philosophical opinions . . . take their chance in the world."

"Poor Buissart!" exclaims historian Jessica Riskin. "The claim that lightning rods attract lightning to houses that would not otherwise have been struck was at the heart of his opponents' case. In response, he could only cite the leading authorities on lightning rods, who offered a resounding yes, no, and maybe."

In May 1783 the case was heard in Arras, the provincial capital of Artois. Buissart, having written the extensive brief, now handed the job of arguing the matter in court to a fiery protégé, Maximilien de Robespierre, a twenty-five-year-old attorney known less for his command of the law than for his eloquence. "The mind of Robespierre was made for such a case," a popular biography of the future revolutionist explains. "Here was philosophy and all the light of the century called into question! *Here was Franklin to be defended!*" Robespierre was comfortable arguing a scientific subject; as a student, he'd shown so keen an interest in meteorology his schoolmates had nicknamed him "Barometer."

By coincidence, Vissery's trial took place in summer 1783, a strange and distressing season for contemplating man's relationship to nature. It was a summer of strong atmospheric electricity and a permanent dry fog over much of Europe. Days seemed hot and airless, the nights unusually cold. Even on clear days the sun's light was opaque, and diffused by haze, and on most nights the moon and stars were invisible. The air itself seemed fouled, and respiratory illnesses struck down children and the elderly. As the naturalist Gilbert White of Selborne noted, "there was reason for the most enlightened person to be apprehensive."

It was Franklin who first theorized correctly the source of the problem. Recently, the world had experienced four major volcanic eruptions. The first had occurred in December 1782 on the Japanese island of Honshu, another elsewhere in Japan in April 1783, and then two more near Iceland in May and June. Franklin suggested that the "dry fog" and the perturbations of light and temperature were a result of ash and dust from these eruptions lingering in the upper atmosphere. Franklin, in time, was ascertained to have been right, but while the eerie conditions persisted, no explanation seemed adequate to quell the public's unease.

Robespierre made two speeches before the court in Arras. Interestingly, he abandoned the position Buissart had crafted in his plea—that the aldermen had erred by not consulting experts—and made a far different argument: He insisted that the judges of the Conseil d'Artois, men of the law, were perfectly qualified to judge lightning rods. Humanity's knowledge of lightning was based, after all, he said, on our daily experience. Experts might be required to rule on intricate theories of physics, but anyone with common sense could recognize simple scientific fact. Thus the aldermen of Saint-Omer had erred not by neglecting to call in experts but by ignoring the wisdom of their own observations and experience.

Robespierre's case centered around several incontrovertible facts—that electricity is attracted to metal, that it will strike and conduct itself by metal, and that the lightning rod functions by tak-

ing advantage of these certainties. "Lightning accepted its laws," Robespierre said of the rod, "and thereby immediately lost this blind and irresistible impulse to strike, smash, overturn, crush all that stands in its way. It has learned to recognize the objects it is to spare." Robespierre was willing to admit that some aspects of the science remained unknown or inexact, but he insisted they were irrelevant, since experience had already proven the efficacy of lightning rods. He drew a comparison to the once controversial "art" of inoculation, saying, "We must calculate the victims art has saved and those nature has sacrificed. . . . Because this calculation generally proved that men gain more from confiding themselves to art than to nature, inoculation has triumphed over all obstacles." Emphasizing the lightning rod's simple utility, Robespierre mocked the poetic idea of Benjamin Franklin controlling thunderbolts as an image equally as ludicrous as God hurling them.

Is man so unfamiliar with miracles that another prodigy leaves him stupefied? Have the Sciences produced for his benefit few miracles that he must consider this new boon beyond his power? . . . There is no miracle here. That man has dared wrest the lightning from heaven; that he controls all its movements however he chooses; that he says to it: be careful not to touch these buildings; come, follow the route I have set for you, and hasten to bury yourself in the abyss I have made for you; there is a prodigy: but it is also nothing more than a product of the imagination.

Ironically, in order to counter Robespierre, the prosecutor was forced to assume Buissart's original position that experts were required to pass judgment on lightning rods. One authority cited was Jean-Paul Marat, the physician and later revolutionary journalist, who had written substantially on the subject of electricity. Marat had observed that lightning rods were not always capable of drawing the electricity from a large cloud, and claimed there were examples of rods being blasted to pieces by lightning. But Marat's scientific expertise was not widely respected. Even as statutes were being enacted across Europe outlawing the ringing of church bells during a storm, he had written to suggest that churches retain their ancient method of ringing bells when a thunderstorm approached, if only as an added precaution to having a lightning rod.

Robespierre's appeal to the common sense and utility of the lightning rod carried the day, and Vissery won his case. Described as a "young lawyer of rare merit" by the Parisian press, the upstart moved swiftly to reap some additional benefit by pressuring Vissery to underwrite the costs of publishing his pleas before the court. A grateful Vissery did so, and the text appeared or was excerpted in numerous periodicals, and was sold in pamphlet form. Robespierre sent a copy to Franklin at Passy:

Sir, A writ of condemnation by the magistrates of Saint-Omer against electrical conductors furnished me the opportunity to appear before the Council of Artois and plead the cause of a sublime discovery for which the human race owes you its thanks. . . . I dare hope, Sir, that you will deign to have the goodness to accept a copy of this work, the object of which was to induce my fellow citizens to accept one of your gifts: happy to have been useful to my country by persuading its first magistrates to welcome this important discovery; happier still that I can add to this advantage that of being honored by the good graces of a man whose least merit is to be the most illustrious scientist of the universe.

If Franklin made any response, it is nowhere recorded. As for Vissery, back home in Saint-Omer, his taste of victory quickly soured. Clambering onto his rooftop to restore the large sword atop his lightning rod, he was pelted by rocks and brickbats hurled from the street below, and exchanged curses with his neighbors. When he died the next summer, he left explicit instructions that whoever should live in his house should maintain the rod, as it represented Progress, and its right to exist had been so nobly defended. The new owner, however, happening to be one of the aldermen who had originally sat in judgment of the lightning rod, soon arranged for its removal.

The Saint-Omer case of 1783 saw new technology challenged by a wary public, with both sides turning to the law as arbiter. The following year, another controversy arose in France: This time it was the turn of science to question a dubious healing faith held dear by much of the public. While Benjamin Franklin had been present in the Saint-Omer dispute only through his theories of lightning protection, he would assume a central role when the French scientific establishment confronted one of the late Enlightenment's most enigmatic figures, the healer Franz Anton Mesmer. Franklin, who had had his own "heretical rod" attacked by skeptics, was now asked to lead the assault on the "mesmerizing rod" of an arriviste whose success in France compared to his own, and whose powers were acclaimed by no less than the Marquis de Lafayette and members of the French royal family.

Throughout the eighteenth century, natural philosophers had enjoyed a privileged relationship with the public, astonishing them with experiments that brought electricity into one's parlor and lifted balloons into the unknown region above the clouds. Inherent in such a relationship was the possibility that the public—"impressed by the marvelous" but "devoid of the faculty of discernment"— would grow dangerously credulous. Much as New Englanders of the seventeenth century dwelled in "an enchanted universe" of providences and supernatural signs, a hundred years later it was science that made remarkable things come true.

Inevitably, this openness to what had previously seemed impossible led to fraud and outright hoaxing. Charlatans fed the appetite for demonstrations and spectacle that the natural philosophers had created, as did a number of mostly harmless scientific myths. It was said that sperm, viewed through a microscope, contained a tiny fully formed man; that nature had split fruits like the orange into convenient sections so that a family might more easily share them; and that an actual flame burned within the human heart. One of the more effective hoaxes was perpetrated by a London commoner named Mary Tofts, who briefly convinced everybody, including the royal physician, that she could give birth to rabbits. Perhaps the most celebrated charlatan was the spiritualist Alessandro Conte di Cagliostro, who claimed to have known Jesus Christ and boasted of the ability to set seawater aflame. Dividing his time between London and Paris in the 1770s and 1780s, he led séances in which the spirits of departed nobility and *philosophes* were summoned, before venturing to Rome where he was imprisoned for heresy and what the Vatican deemed his "titanic impostures."

The influence of these pretenders concerned the established scientists of the academies and royal societies, but most were soon exposed or could be easily ignored. What complicated the case of Franz Mesmer was that he was far more refined than the coarse Cagliostro, and unlike an outright sham artist such as Mary Tofts, seemed to sincerely believe his own claims. By bringing forward a system of healing that invoked a subtle fluid (magnetism) and that satisfied the era's insistence on truth as experience (since his patients claimed to be cured), he successfully harnessed mainstream

Enlightenment thought, creating a theory and a methodology that could not be readily demolished.

Mesmer was, by all accounts, charismatic as well as handsome. One biographer describes him as "imposing of mien, penetrating of eye, imperial of forehead, Roman of nose . . . [and] deep of understanding." A native of southern Germany, he became convinced, while a university student, that magnets derived their attractive and repulsive powers from the stars, and that the sun and moon influenced the functioning of human health just as they did the tides. Opening a medical practice in Vienna in 1768, he soon did away with the moon and sun, and then the magnets, ultimately claiming himself physically capable of directing a magnetic healing force. Patients drawn to Mesmer's clinic complained of paralysis, eye irritations, internal blockages of the intestine or menstrual tracts, or neurological symptoms consistent with anxiety and depression. He taught, however, that there was in reality only one human ailment, the blockage of magnetic fluid. Mesmer's treatments consisted of his touching patients sympathetically with his hands, an iron rod, or a wand, until a brief convulsive trauma in the patient occurred as the "obstruction" was confronted. Health returned as the body responded smoothly to the renewed flow of natural fluid, which one patient described as "a warm wind."

It was perhaps no wonder that people with long-term or incurable conditions would place their faith in such a promising treatment: Compared with bleeding, purges, amputation, crude electric shocks, and other methods that passed for progressive medicine in the eighteenth century, Mesmer's "warm wind" must have seemed exceedingly humane. Moreover, in treating depression, anxiety, and other physiological and psychological ailments, Mesmer was addressing problems traditional medicine had not yet touched. "[The] aggressively rationalistic imperatives of the epoch," notes scholar Terry Castle, "also produced, like a kind of toxic side-effect, a new human experience of strangeness, anxiety, bafflement, and intellectual impasse." Mesmer, whose emphasis was on physical healing, had, without intending to, and in anticipation of modern psychology, hit upon the vital dynamic of the troubled unconscious.

Franklin's musical invention, the armonica, had a central role in Mesmer's sessions. "Do you know that Herr von Mesmer plays Miss [Marianne] Davies' armonica unusually well?" Leopold Mozart wrote to his wife from Vienna in 1773. "He is the only person in Vienna who has learnt it and he possesses a much finer glass instrument than Miss Davies does. Wolfgang too has played upon it. How I should like to have one!" Wolfgang Amadeus Mozart wrote a quintet for armonica for Marianne Kirchgessner, used Mesmer's house to stage a recital based on one of his operas, and gratefully included a scene about mesmerism in *Così fan tutte*.

Mesmer's links to the world of Viennese music, however, would also lead to scandal. One of his patients was Maria Theresa von Paradies, a gifted eighteen-year-old pianist, blind since early childhood, whose recitals Mozart had attended and whose playing had so moved her godmother, the empress Maria Theresa, that a lifetime state pension had been bestowed upon her. The girl, who had been subjected to electric shocks, suffered from fits of rage and "melancholia." In 1777, Mesmer, prescribing an intensive course of treatment, moved Paradies into his house, where two other young female patients already resided. This arrangement soon worried Paradies's parents, who feared Mesmer had usurped their authority and was perhaps taking other liberties. Franklin's friend Jan Ingenhousz, the Dutch physician now residing in Vienna, had once been happily disposed toward Mesmer, but had recently become a critic, and appears to have played a role in encouraging the parental concern. When they demanded their daughter's release, Mesmer claimed his treatments had restored her eyesight, and that she was now feigning blindness to safeguard her royal pension. The highly questionable situation was resolved amid rumors that Mesmer had seduced the girl, providing grist for those within the local medical establishment already skeptical of his methods. With his reputation tarnished, Mesmer departed Vienna.

Arriving in February 1778 in Paris, Mesmer opened two healing clinics and sought recognition from France's leading physicians and scientists. Several, including Franklin's physicist friend Jean Baptiste Le Roy; Felix Vicq d'Azyr, perpetual secretary of the Society of Medicine; and the chemist Claude-Louis Berthollet, agreed to hear Mesmer out. However, when Mesmer appeared before the Academy of Sciences, attendees reacted with derision, and at a subsequent demonstration, many walked out. More damaging to Mesmer was a hoax perpetrated by Antoine Portal, a surgeon, who sought treatment for a feigned ailment at one of Mesmer's sessions and then regaled the rest of Paris with his account of an evening spent amid such obvious foolishness.

Franklin agreed to meet with Mesmer, but the latter went away frustrated because the Philadelphian was far more interested in discussing the armonica than Mesmer's theories of healing. Franklin may have only been trying to be polite, for he was likely dubious about mesmerism to begin with, although, like Mesmer and Mozart, he did subscribe to the idea that "glass music" was capable of promoting human harmony. Still, there can be little doubt that Mesmer, precisely because the ethereal sound of Franklin's armonica blended so ideally with his own healing techniques, hoped for greater sympathy from the instrument's inventor, as well as some kind of endorsement.

Even as acceptance by France's scientific community eluded him, Mesmer's practice in Paris became so popular he was forced to adapt his treatments to accommodate ever larger groups of patients. For these "séances" he devised a *baquet*, a long oaken trough about a foot and a half deep in which glass and iron filings soaked amid carefully arranged bottles. The lid of the *baquet* had holes through which as many as thirty iron rods were inserted, with patients sitting by its side, each holding one of the rods. The patients were joined together by a sash, or clasped one another's thumbs as a means of increasing the tub's healing power. The *baquet* thus became a kind of mesmeric "Leyden jar," a storage battery capable of restoring an equilibrium of magnetic fluid. The patients "recharged" themselves by touching the rods to the regions Mesmer had diagnosed as troubled, usually their head, nose, or abdomen. Mesmer, dressed in lilac taffeta, then went among them, touching and counseling, probing his patients' bodies with his fingers or a wand, while an armonica droned ambiently in the background. Participants who went into seizure were helped or carried by Mesmer's valet, Antoine, into an adjoining mattress-lined "crisis room." When the number of patients became too large for his clinics, Mesmer reconceived his treatment sessions for use outdoors, where he magnetized trees to replace the *baquet*, and roped his eager patients to them.

The reaction of Franklin or other philosophers to Mesmer's healing fluid could not help but be colored by another controversial healing doctrine, the use of medical electricity. When electricity first became popular in Europe in the 1740s, the sparks and snaps of Gray's dangling boy or the kisses of an electric Venus made for compelling entertainment. But anyone who touched a Leyden jar or took part in a "circle charge" understood immediately electricity's unique power to instantly pervade the entire body. It was no great leap then to wonder if this universal force might possess the capability of curing disease or paralysis, or of providing other healing. And because the times were so desperate for promising medical treatments of any kind, electricity produced high expectations, its touted cures ranging from bed-wetting to infertility.

Natural philosophers had always wanted to believe that electricity had a grand purpose. Franklin with his lightning rod had shown one practical application of man's *knowledge* of electricity, but surely a use for electricity itself must exist. Prior to the invention of the electromagnetic telegraph in the nineteenth century, medical cures appeared to be the new science's most likely application, and a devoted core of medical practitioners, journalists, the public, and of course the usual opportunistic quacks propped up and reinforced one another's expectations.

Interest in medical electricity was always strongest among paralysis cases, as it seemed almost intuitive that "sleeping" limbs might be awakened by electric shocks and thus restored to life. In the wisdom of the day, and as per Mesmer, loss of sensation, numbness, and the immobility of limbs were believed to be caused by blockages of natural "life" fluids or processes; electricity, which was itself thought to be an elemental life force, would, it was believed, clear such obstructions and reintroduce an even flow of vitality. Those doctors who reported successful electric cures of paralysis were often besieged, hopeful patients appearing from as far as a hundred miles away, hobbling on crutches, strapped onto crude litters, or carried by friends and family—pilgrims on a quest of faith.

Franklin generally appeared satisfied to write about and advocate for progress in medicine; he was not, it seems, an eager practitioner. In Philadelphia he had applied electric shocks to "palsies"—out of scientific curiosity at first, and later, it seems, more to indulge those who came to him, desperate for any possible cure. Even the venerable James Logan, who was paralyzed by a stroke and whose health deteriorated in the 1740s, turned to electric shocks, which were administered either by Franklin or by his friend Lewis Evans. To those who came to Franklin for help, he gave fairly powerful shocks from a battery of two six-gallon Leyden jars. His patients reported a sensation of warmth from the treatment, pricking sensations during the night, and stricken limbs that often did *seem* to regain strength and mobility. But Franklin found that after several days, the positive therapy diminished and its subjects went home discouraged.

"I never knew any advantage from electricity in palsies that was permanent," Franklin wrote in a 1757 letter to John Pringle. He said he thought the benefit people felt had more to do with "the exercise in the patients' journey, and coming daily to my house," and the "spirits given by the hope of success, enabling them to exert more strength in moving their limbs." Although he concluded that the patients' own hopes had created their momentary "recovery," he later allowed that the shocks he administered might have been too severe, and that perhaps a series of far lighter shocks would bring better results. Indeed, over the years, as medical electricity became more refined, a series of very mild shocks came to be preferred to a few strong ones, often with some medicinal powder mixed into the water of the Leyden jar so that its curative effects might enter the patient's body along with the electricity. In Franklin's view, curing paralysis may have been beyond medical electricity's powers, although as late as 1785 he was agreeing with his friend Jan Ingenhousz that electricity could someday become useful in treating the insane.

Franklin did take an interest in the curative powers of electric fish—most notably the *Torpedo marmorata*, or crampfish, known to give powerful shocks from organs located on the side of its head. The shocks instantly incapacitate prey by producing in victims a substantial numbness, or *torpor*. The torpedo's strange powers were recorded in the time of Aristotle, and there was a persistent belief that they cured gout, an inflammation of the joints, which perhaps was what originally helped pique Franklin's curiosity, since he suffered from the disease. In first century A.D. Rome, the physician to Emperor Claudius recommended that gout sufferers travel to the seashore, find a live torpedo washed up by the sea, and stand on top of it until their afflicted leg was numb up to the knee. Franklin closely followed the research carried out by John Walsh, a natural philosopher and member of the House of Commons, who described the torpedo to Franklin as "an instinctive electrician," one whose powers lay beyond "that veil of nature, which man cannot remove." Walsh found that the fish was able to recharge itself instantly, as he observed it to give as many as twenty shocks within a minute when agitated. What most intrigued Walsh and Franklin was that the torpedo gave a shock as from a Leyden jar, but without any spark and snap. In Franklin's day, before the concept of electrical current was known, this presented a mystery, one that both Franklin and Walsh were forced to leave unsolved.

The thrall for medical electricity persisted despite the lack of convincing studies supporting it, and even despite frequent debunking. Nollet, an early believer in the medical applications of the Leyden jar, had by the late 1740s become concerned that he could not replicate the successful treatments being reported by Italian electricians. In summer 1749 he journeyed to Italy at the expense of the French government to see if perhaps the specific "atmosphere" of that country was contributing to these cures. The Italians were unable to demonstrate electric cures convincingly, and Nollet returned to France questioning medical electricity generally and what he called its "temporary shadows." Franklin would later comment favorably on Nollet's candor in investigating the Italian claims, as would others.

As Franklin's experience with his "patients" suggested, one factor complicating any evaluation of medical electricity's effects was the inadvertent collusion of those being treated. As he once confided to his friend Benjamin Rush, "quacks are the greatest lyars in the world, except their patients." In the few controlled studies of medical electricity that were performed in Franklin's day, the participants tended to be peasants, workmen, or maids who, out of deference to the gentlemen experimenting on them, were seemingly inclined to report just those effects that the investigator was seeking. Tests of medical electricity's efficacy thus lacked the credibility even of Franklin's kite or sentry box demonstrations, in which natural philosophers themselves physically ascertained the results. Still, electrical medicine held far too much appeal to be entirely dismissed, and its quackery was, after all, often indistinguishable from what passed for more formal medical treatment. Testimonials to the effectiveness of medical electricity, frequent announcements of innovative new methods, and much boosterish writing on the subject continued apace throughout the late eighteenth century, despite the absence of confirming data.

For five years, mesmerism thrived in Paris and spread to the rest of France. A *Société de l'Harmonie Universelle* was founded to provide training in the mesmeric arts to a group of disciples, with branches soon operating in Lyon, Marseilles, Bordeaux, Nantes, and elsewhere, bringing Franz Mesmer lucrative fees and even greater fame. He lived comfortably in a hotel in Paris's fashionable Marais, and traversed the city in a stylish carriage. As in Vienna, however, a rumor that he was taking advantage of vulnerable women who had given themselves into his care began to arouse opposition. Often, Mesmer treated his patients by sitting directly across from them with knees touching. He used his hands, which he regarded as two magnetic poles, to massage patients gently in the area of their affliction so as to draw the healing fluid from hand to hand and through their body. It was the resulting gossip that, as Claude-Anne Lopez suggests, mesmerism was not only medically unsound, but perhaps also morally so, that ultimately led to an official inquiry.

The call for the inquiry originated with Dr. Charles d'Eslon, physician to the Comte d'Artois and the most prominent Frenchman to have been won over to Mesmer's methods. Several of Mesmer's Harmony Society disciples, including d'Eslon, were beginning to chafe under his control; they had served their apprenticeships and wanted to be regarded, and compensated, as full-fledged healers. This threatened Mesmer, who now loudly claimed that he and only he possessed the secret power that made the healing effective. For years, d'Eslon had been a key ally and diplomat for Mesmer, winning a degree of indulgence from the French authorities (Mesmer barely spoke French) and helping the newcomer avoid controversy. Now d'Eslon, still confident in Mesmer's techniques but hurt by Mesmer's prevarication, and concerned that his own name might be besmirched by rumors of immorality and charlatanism, looked to a fair inquiry to establish the practice's legitimacy once and for all.

In March 1784, a joint investigative committee was formed from members of the Academy of Sciences, with Franklin as one of the co-chairmen, along with the astronomer Jean-Sylvain Bailly, later mayor of Paris; Dr. Joseph-Ignace Guillotin, known for having recommended a humane beheading device for executing criminals; and the pioneering chemist Antoine-Laurent Lavoisier. (Lavoisier had already won fame for helping debunk a scientific assumption even more prevalent than mesmerism, the theory that a weightless, odorless substance known as phlogiston existed within all physical matter and gave off heat and energy when it was consumed. In 1779, following up on work by Priestley, he confirmed the fallacy of this theory and declared a ubiquitous gas, which he named oxygen, to be the active agent.) The next month, the monarchy appointed a second body of inquiry from members of the Royal Society of Medicine, although it was Franklin's group that was seen as the lead investigators. The inquiry concentrated officially on d'Eslon's healing methods, as Mesmer had made it clear he would not cooperate with either of the commissions; but it was obvious that mesmerism itself was now on trial. "Never has a more extraordinary question divided the minds of an enlightened nation," the Franklin committee wrote in appraisal of its own task, in what appears to be history's first official inquiry of alleged medical or scientific fraud.

While Franklin and his colleagues may have thought Mesmer sincere in his beliefs, they suspected mesmerism of being, if not a gross deception, a false science at best. Franklin likely knew that at home in America, Mesmer and his flaky entourage—d'Eslon, the valet Antoine, and the others—would have long since been run out of town. But hasty pronouncements from his committee might only serve to embolden Mesmer's defenders, and he could, after all, sympathize with those who wanted so badly to believe in Mesmer's miracles; he knew from his own experience that even a conjured or imagined cure was often of tremendous value to a patient. "There are in every great rich City, a Number of Persons who are never in health, because they are fond of Medicines, and always taking them whereby they derange the natural Functions, and hurt their Constitutions," he said. "If these people can be persuaded to forbear their Drugs in Expectation of being cured by one of the Physician's Fingers or an Iron Rod pointing at them, they may possibly find good Effects tho' they mistake the Cause."

Franklin proceeded carefully for personal and diplomatic reasons, as well. His grandson William Temple Franklin, who served as his secretary in Paris and who was so often in his *grand-père*'s company the French dubbed him "Franklinet," happened to belong to one of Mesmer's Harmony Societies, a fact Mesmer himself made sure was known. And there was something inherently delicate about an American, even one as esteemed as Franklin, adjudicating a scandal that centered on the apparent gullibility of the French public. After all, as Joseph Michel Antoine Servan, one of Mesmer's supporters, asked, was it more credulous for Franklin to claim the control of thunder than for Mesmer to describe a magnetic fluid with curative powers?

From the depths of America, an almost unknown land, a man even more unknown than his country, stood up to cry: "men, listen to me! I have the power to draw thunder from the sky, and I can often force it to fall upon any point on earth it pleases me to choose:" What mockery from one pole

to the other!... Franklin... you would have been condemned to eat crow, and to abandon right then and there
your physics and your genius.

Franklin, of course, had been challenged for his advocacy of
lightning rods, although rarely scoffed at or ridiculed, as Mesmer
had been. And his science had for the most part been verified. But
just as Franklin's friends might proffer examples of lightning rods
successfully sparing life and property, didn't legions of happy patients testify to mesmerism's salutary effects? Mesmer's ideas, it
could be said, were not mere hypotheses; they produced observable results. Considered by natural philosophy's empirical standard, Claude-Anne Lopez points out, Mesmer's "magnetic fluid"
was no more fantastic than Newton's gravity, Franklin's electricity,
or the balloons of the Montgolfiers. Mesmer said he wielded the
power of a subtle healing fluid, and there was abundant evidence
that the men and women who swooned and convulsed and went
away from his séances "cured" had experienced it.

The investigators began by submitting themselves repeatedly
to mesmeric treatment around a *baquet*. Feeling no effects, they
reached their first conclusion: that anyone who doubted the healing
method's efficacy—who did not believe in it—would not experience its results. The greater challenge was to isolate what Mesmer's
fluid actually gave his patients, and what caused them to feel such
strong beneficial sensations. To do this, Franklin and his colleagues
introduced, likely for the first time in medical history, a placebo. In
Franklin's garden at Passy, the committee members impersonated
Mesmer or d'Eslon, and watched as blindfolded patients expressed
joy at feeling the fluid's effects; but in reality there could be none,
since the committeemen were only mimicking the mesmerists'
words and gestures. They then had actual mesmeric healers "mag-
netize" subjects without the latter's knowledge, and observed that
the fluid had no effect. In one of the most dramatic experiments, a
twelve-year-old boy known for being especially responsive to the
procedure was selected, and while he was kept inside Franklin's
house, d'Eslon magnetized a single tree outside in the garden. The
boy was then blindfolded and told to locate it. After embracing sev-

eral trees that had not been magnetized, he went into convulsions and passed out. As the commission's report noted:

> If the young man had felt nothing even under the magne-
> tized tree, one could simply say that he was not very sensi-
> tive—at least on that day. But the young man fell into a crisis
> under a nonmagnetized tree. Consequently, this is an effect
> which does not have an exterior, physical cause, but could
> only have been produced by the imagination.

The investigators, concluding that no magnetic fluid appeared to exist and that any effects produced must result from the patients' imagination, their sensitivity to touch, or their imitation of others, denounced the healing practice in a widely published report to the king. Noting that Mesmer and his magic wand "seemed to transport us to the age and the reign of the fairies," the commissioners said that the ascribed powers of mesmerism did not work on those who did not expect to be affected, and that "while proceeding . . . we witnessed the disappearance, one after another, of the properties attributed to this supposed fluid, and that the whole theoretical construction, built on an imaginary basis, crumbled before our eyes." A confidential side report that went only to Versailles warned explicitly that women, because of what Franklin and his colleagues assumed to be their more emotional constitutions, were the natural victims of such a fraud, and that, since mesmeric patients were often women and the mesmerists were always men, involvement in Mesmer's Harmony Societies placed them at special risk.

Before ending, the committeemen suggested that while magnetism did not work physically to heal people, it nonetheless raised issues pertaining to "a science which is still brand new, the science of the influence of the psychological on the physical." They cautioned that the power of imitation among people in a group and the tremendous vigor of human imagination, while potentially creative forces in society, might also bring it to ruin. The sudden mixture of imagination and imitation in a crowd, Franklin and his co-authors warned, could become volatile, firing men to act on their most base passions.

If not read by all Parisians, the report's conclusions were powerfully communicated by a popular engraving, *"Le Magnétisme Devoilé,"* that depicted the committeemen, with Franklin in the lead holding their scrolled report aloft, barging into a Mesmer séance, bringing sunlight into the murkiness and knocking over a *baquet,* as Mesmer and d'Eslon flee, one on a broom, the other aboard a flying donkey. For d'Eslon, who had looked to the inquiry to safeguard his reputation, the combined impact of the committee's report and *"Le Magnétisme Devoilé"* could not have been more devastating, and he died soon after, reportedly while being magnetized. Mesmer left Paris in 1785 and, after several years roaming Europe, visiting his remaining Harmony Societies and defending his methods, he retired quietly to a village on Lake Constance in southern Germany. His last request at his death in 1815 was that the priest attending him play an air on the armonica.

Franklin also left Paris in 1785 and headed home to Philadelphia, now the capital of the United States of America. The world had turned many times in the nine years he had lived in France, as he watched from afar the fortunes of the Continental Army and cheered its ultimate victory at Yorktown, achieved with the help of the French naval and military forces whose alliance he had done so much to secure.

His departure from France was one long sad celebration of farewell, for he was soon to be eighty years old and his friends knew he would not return again. "If I had no country of my own," he assured them, "it would be at Paris where I should like to finish out my days; but I want to enjoy for a moment the pleasure of seeing my fellow citizens free and ready for all the happiness I wish them." At Passy he was seen off by sobbing admirers who entreated him to reconsider, even as he and his luggage were loaded aboard a royal litter led by two large mules loaned by Marie Antoinette. Some close friends accompanied him to Le Havre, others went along as far as Normandy, and one crossed the channel with him to Southampton. There, English acquaintances and friends from the realms of science and politics came down from London to pay their respects—the Old World seeing him off one last time, in a slow and deliberate leave-taking.

As he was about to leave France, Franklin had received an unexpected visitor. Martinus Van Marum, a young electrical experimenter from the Teyler Museum in Haarlem, Holland, was a staunch Franklinist who had written on the single-fluid theory. The museum had the world's largest electrical friction machine, designed by Van Marum himself, with rotating discs sixty-five inches in diameter (the previous largest were eighteen inches). The device required as many as four assistants to crank it into operation, and could create sparks two feet long between its large brass electrodes. The flash it produced so convincingly represented a bolt of lightning that Van Marum utilized the machine to observe the melting rates of various metals used in the manufacture of lightning rods. These sparks, with their huge veinlike branches clearly visible as they shot between the conductors, were able to register on an electrometer positioned forty feet away, and were far easier to study than those from smaller machines.

Van Marum visited Paris in June 1785 and shared with members of the Academy of Sciences, including Le Roy and Lavoisier, a printed impression the Teylerian machine had made of the electrical charge as it traveled from a positive to a negative conductor. The print showed the pattern of a powerful spark discharge that indicated a unidirectional flow, rather than two fluids flowing in opposite directions.

As Franklin was then still at Passy, Van Marum was encouraged to show him the image. At first he was timid about approaching so great an eminence, but Le Roy offered to go as his companion. They arrived as Franklin was busy packing.

> On being first announced, we perceived that it was his intention, after having granted us a short interview, to excuse himself on account of his approaching departure. But perceiving that the experiments, of which I offered him the descriptions, were very important with regard to that branch of physics in which he himself had formerly made so many

successful researches, this venerable old man, whose presence inspired me with such profound respect, made me sit
down beside him, and begged me to communicate and
point out to him whatever I judged most likely to throw a
light upon this science.

Franklin, peering through the lower half of his bifocals, closely
studied the image, looking back and forth between the opposite
conductors. He asked Van Marum if it showed accurately the electrical exchange that had taken place, and when the younger man
affirmed that it did, Franklin, much pleased, exclaimed, "This
then proves my theory of a single electric fluid, and it is now high
time to reject the theory of two sorts of fluids." He then excused
himself to prepare for his trip home.

The westward journey, lasting from July 28 to September 14,
was for Franklin a kind of imposed vacation from politics and the
affairs of the world, a rare interlude that only an ocean voyage
could provide. He had booked an extra-large cabin on the chance
that some friends from London might be joining him, but when
they decided against it he retained the cabin and, as the trip was
largely calm, had ample room as well as free time in which to write
and make observations. Neglecting the advice of those who had
encouraged him to use the journey to work on his autobiography,
he instead spent the voyage indulging, to great extent and for the
last time, his lifelong fascination with the sea. He produced two
excellent papers in the form of letters to scientific friends: one a
lengthy missive to Ingenhousz, "On the Causes and Cure of
Smoky Chimneys," and another, "Maritime Observations," addressed to Alphonsus Le Roy, a brother of Franklin's longtime
Parisian friend.

The latter paper is the more illuminating and personal, a compendium of his thoughts about the mysteries of the ocean and the
ships that sail upon it. Growing up amid the sights and sounds of
maritime Boston, he had as a boy dreamed of running away to sea;
on his first return voyage to America in 1726, he wrote extensively
in his journal of dolphins, weeds, sharks, birds, shellfish, and a
lunar eclipse, initiating a lifelong practice of making scientific ob-

servations during his transatlantic voyages. Fifty years later, during the wintry 1776 voyage to France, biographer Carl Van Doren writes, "The indomitable old man, who was almost certain to be hanged for high treason if the *Reprisal* should be captured, noted the temperature of air and water every day." Even when his affairs kept him on terra firma for long periods of time, he nonetheless pursued maritime interests, such as his active sponsorship in the 1750s of a Philadelphia-based expedition to locate the Northwest Passage.

On this, his final ocean voyage, Franklin was determined to test a phenomenon he had wondered about for years: the warm northward current known as the Gulf Stream. As early as 1746, Franklin wrote to Cadwallader Colden to ask if he could answer the riddle of why ships sailing east toward England complete their voyage more quickly than ships sailing the exact route westward. In 1768, the Board of Customs at Boston asked the British Treasury in London why British packets from Falmouth, England, to New York took two weeks longer than American merchant vessels that sailed from London to Rhode Island. The Treasury put the question to Franklin, who in turn wrote to a kinsman on his mother's side, Timothy Folger of Nantucket. Folger, a veteran mariner, explained that the American vessels had a swifter crossing because their captains were familiar with the Gulf Stream, adding that Nantucket whalers probably knew the stream best because whales hung by its sides to feed on the plankton borne off by the flow of warm water. Folger mentioned to Franklin that the whalers had often told British vessels they encountered on the high seas about ways to avoid struggling against the Gulf Stream, but that such advice had generally been disregarded. "We have informed them that they were stemming a current that was against them to the value of three miles an hour and advised them to cross it, but they were too wise to be counseled by simple American fishermen!"

With Folger's help, Franklin obtained an idea of the Stream's boundaries and produced a basic chart. "Discoursing with Captain Folger," Franklin informed the secretary of the British Post Office, "I received . . . the following information, viz.: . . . that the whales are found generally near the edges of the Gulph Stream, a strong

current so called, which comes out of the Gulph of Florida, passing
north-easterly along the coast of America, and then turning off
most easterly, running at the rate of 4, 3½ and 2½ miles an hour."
Franklin explained that westbound ships hesitated to cross the
stream because they were fearful of encountering the shoals near
the North American land mass, but that as a result they often remained heading south in the Gulf Stream and losing valuable time,
often fifty to seventy miles per day, sometimes even getting carried backward toward Europe.

As Franklin correctly understood, the origins of the Gulf Stream
are in the shallow Gulf of Florida, where water is warmed and then
shot through the Straits of Florida. The narrowness of the Straits
increases the speed of the current, so by the time the flow enters
the Atlantic again and heads northward, it is moving faster than the
surrounding water and is considerably warmer, at times by as much
as 20 degrees. Off Newfoundland, the warm water encounters the
chill air of the Outer Banks, creating the dense fog for which the
Banks are known. Ultimately it turns eastward toward Scandinavia
and Europe.

What was known about this mysterious "river in the ocean"
came from informal accounts of seafarers who, since the days of
Hernando Cortés, had used it as a kind of expressway from the
West Indies back toward Europe. But until Franklin, no one had
synthesized all the available information. While returning to

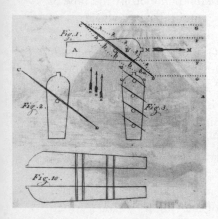

Philadelphia in 1785, he deployed an instrument he had devised to measure water temperatures beyond a depth of one hundred feet, and managed to ascertain both the location and the width of the stream. His measurements have held up well; Franklin's 1786 map agrees in great detail with photographic images of the Gulf Stream made by orbiting spacecraft more than two centuries later.

One nautical issue that Franklin had wrestled with but never completely solved was that of lightning strikes on ships. For engendering a sense of complete helplessness in its victims, nothing in the eighteenth century compared with a storm on the high seas, and high-masted ships on the platform of the ocean's surface presented lightning with an obvious target. A direct hit from a thunderbolt was capable of not merely damaging a ship, but of sinking it altogether, for the parts most likely to be struck, the foremast and mizzenmast, usually passed through the powder magazine. A sudden powder explosion aboard a wooden ship would of course be catastrophic, perhaps leaving little evidence the vessel or its crew had ever existed.

Franklin had waded into the question in 1749 when he forwarded a letter to the *Philosophical Transactions* from a Captain John Waddell. The captain reported that on January 9, 1748, his ship *Dover,* en route to London from New York, encountered a highly charged electrical atmosphere that made "sundry very large Comazants (as we call them) over-head, some of which settled on

the Spintles at the Topmast Heads, which burnt like very large torches." Waddell's experience gave evidence of all that could go wrong when a ship was struck by lightning. Soon after the appearance of the "Comazants," at about nine o'clock in the evening, a deafening blast rent the sky above the *Dover* and lightning split the mainmast, damaging the vessel down to its keel. The impact threw Waddell and his crew to the deck, blinding several temporarily and destroying the magnetic function of the ship's compass. This left Waddell to improvise a three-hundred-mile quest for a safe landing, the damaged vessel limping along with a wounded and frightened crew, no horizon in sight or any reliable means of navigation, until at last the *Dover* stumbled upon the Isle of Wight.

Such hair-raising incidents, reported from every one of the seven oceans, were discussed in tones of awe wherever seagoing men gathered. And eyewitness accounts and officers' reports confirm that there was no subject about which seamen were more superstitious. When lightning struck the seventy-four-gun *Kent* off Toulon, naval officer William Lord Napier had "the dreadful opportunity of witnessing, with my eyes fixed upon them at the moment, not less than fifteen most valuable men, all upon the bowsprit and jib-boom, killed or dreadfully scorched, as it were in the 'twinkling of an eye' . . . Some were precipitated into the water, and others, lying dead across the boom, continued in the posture they had assumed before the accident took place." The *Wasp*, near the coast of Florida, was struck by "extremely vivid and forked" lightning, the crash so deafening that officers and able seamen alike threw themselves to the deck in fright. The faces of two men killed outright by the flash were so horribly blackened and disfigured that the crew begged the captain to immediately commit the dead to the sea.

Franklin, in a letter to Collinson in 1751, noted that the "Comazants" Captain Waddell had reported on his mastheads were "the Electrical Fire . . . drawing off, as by Points, from the Cloud, the largeness of the Flame, betokening the great quantity of Electricity in the Clouds." Franklin proposed that "had there been a good Wire communication from the Spintle Heads to the Sea, that could have conducted more freely than Tarred Ropes, or Masts of Turpentine Wood, I Imagine, there would either have been no

Stroke, or, if a Stroke, the Wire would have conducted it all into the Sea without Damage to the Ship." William Watson had made a similar suggestion in 1762, although Watson's efforts inadvertently wound up convincing the Royal Navy of the impracticality of lightning rods on ships. Watson's well-meant recommendations were for a kind of lightning rod "kit"—long links of copper rod rigged with hemp, to be stored aboard ship and quickly assembled at the approach of a storm. Franklin endorsed a commercially sold kit of this kind in private correspondence in the early 1770s, and again in his comprehensive roundup of maritime issues in 1785; however, in regular use the method proved disastrous, as the apparatus could not be set up rapidly enough in the face of an approaching storm, particularly in rough seas. Sailors responsible for the last-minute rigging of the mechanism, hurrying to drape the links over the masts and down over the ship's side into the water as a storm came on, frequently became the lightning's primary victims, much as bell-ringers in church towers had so often been martyred.

The solution to this dilemma would eventually come from William Snow Harris, a physician and advocate of lightning protection who researched hundreds of British navy reports and was the first to suggest that lightning was likely to blame for some instances in which ships were lost at sea without a trace. Harris recommended making ships' masts themselves into lightning rods, lining them with copper plates that would be connected through the deck to the "principal metallic masses in the hull." This would alleviate the practice of having the crew assemble a portable lightning rod at the last moment. He emphasized that the rod apparatus had to be continuous, and made of copper, the material least likely to fuse under extreme heat, as Van Marum had found in his tests at the Teyler Museum. The rod would be permanently affixed to the ship's mast and would extend through a connecting chain along the hull and into the sea.

There was, unfortunately, great bureaucratic reluctance on the part of the Royal Navy to accept technical reforms in this area because of the trouble with Watson's portable rod and a growing conviction that "land" scientists did not understand shipboard operations. For years, lightning continued its veritable turkey shoot with the mast-tops of Royal Navy and British mercantile shipping—between 1799 and 1815, more than a hundred ships were damaged, and seventy seamen killed—before Harris's reforms were finally adopted in the early nineteenth century.

"I promised to finish my letter with my last observation," Franklin wrote to Le Roy, midway through his paper, "but the garrulity of the old man has got hold of me, and as I may never have another occasion of writing on this subject, I think I may as well now, once and for all, empty my nautical budget, and give you all the thoughts that have in my various long voyages occurred to me relating to navigation." He discussed an idea, credited to the Chinese, of watertight compartments that could be used to keep an injured ship afloat. He condemned the fact that American boats had not been inclined to facilitate this Chinese method, noting that "our seafaring people are brave, despite danger, and reject such precautions of safety, being cowards only in one sense, that of fearing to be afraid." He wrote of the danger of icebergs in the North Atlantic and the movements of ships in fog, and recommended that collisions of ships at sea in foggy weather might be avoided by having someone on deck beat a drum. He advised against eating shipboard poultry, pointing out that ships' cooks were usually ordinary seamen relegated to that duty because of their poor skills as sailors. "God sends meat, and the Devil cooks," Franklin liked to say, having received in his travels his allotment of substandard shipboard fare. He did suggest a design for a soup bowl that would be resistant to spillage and thus would not scald passengers attempting to dine in rough seas. Finally, he qualified his praise for the many improvements that had been made recently in maritime technology by questioning why so many lives and so much ocean traffic were now "employed merely in transporting superfluities" such as tea, sugar, coffee, and tobacco, which "our ancestors did well enough without." He also condemned the seaborne transport of slaves as "the means of augmenting the mass of human misery."

Before leaving France, Franklin had written to Ingenhousz in Vienna to cordially invite him to visit Philadelphia. "Rejoice with me, my dear Friend, that I am once more a Freeman after Fifty Years Service in Public Affairs. And let me know if you will make me happy the little Remainder left me of my Life, by spending Time with me in America. I have Instruments if the Enemy did not destroy them all, and we will make Plenty of Experiments together."

However, even as Franklin, rocking gently homeward across the Atlantic, mused that he might devote his last years to scientific investigation, working contentedly among his bells and Leyden jars, a companion such as Ingenhousz at his side, deep down he likely suspected that things would not turn out that way. For at home, a far different kind of experiment awaited him—the Constitutional Convention of the United States of America.

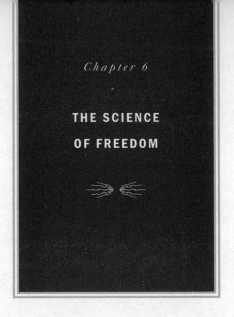

Chapter 6

THE SCIENCE
OF FREEDOM

"PURE AND EXALTED REASON," FRANKLIN'S FRIEND DAVID HARTLEY assured him from London in 1789, "is now rising out of the ashes of ignorance, oppression and slavery, to universal empire in the natural and in the moral world. *Caelo fulmen, sceptrumque tyrannis.*"

Hartley's flattering allusion to the well-known credo of Franklin's fame was altogether fitting. The lightning rod was one of the Enlightenment's greatest inventions not only for the lives and property it saved, but for its potent symbolism. By subduing nature's most arrogant power, it raised a defining question of the late eighteenth century: If reason can vanquish thunderbolts, can it also influence morality, social organization, and human behavior? Had Peter Collinson prophesied correctly that, "Electricity . . . may lead to higher truths"?

"O that moral science were in as fair a Way of Improvement, that Men would cease to be Wolves to one another," Franklin confided to his friend Joseph Priestley in 1780, contrasting the century's rapid advances in natural philosophy with the intractability of human greed and wickedness, "and that human Beings would at length learn what they now improperly call Humanity."

The conviction that a fruitful "science of man" might emulate advances in the physical sciences had been slowly gaining strength for decades. If there had been a flaw in the seventeenth century's scientific revolution of Descartes, Gilbert, Bacon, and Newton, it was that it had been mechanistic, impervious to human feeling. How different the emphasis of the late Enlightenment, when, as Jessica Riskin notes, "knowledge and virtue were inseparable; physics and moral understanding must improve together or not at all, the perfection of each being necessary to the advancement of the other." Newton himself had encouraged his scientific brethren to seek "the very first Cause, which certainly is not mechanical," and asked, "Whence is it that Nature doth nothing in vain; and whence arises all that Order and Beauty which we see in the world?" In his *Opticks*, that veritable manual for experimenters, Newton emphasized that "if natural Philosophy in all its Parts, by pursuing this Method, shall at length be perfected, the Bounds of Moral Philosophy will also be enlarged." Taking Newton's cue two years after the great man's death, John Desaguliers of the Royal Society in 1728 described Newton's cosmos as "the Best Model of Government," and suggested that wisdom and human conscience could serve as a near-gravitational force, one capable of binding humanity together. If providence was no longer assigning social status and individual fate, men were free to act, separately and collectively, on behalf of their own and society's betterment.

This idea hurtled forward through the century, gaining in scope as it traveled. It spread through salons and coffeehouses, in the correspondence of the learned, and in the issuings of Hume, Montesquieu, and the *Encyclopédistes*. "Everything has changed in the physical order; everything must change in the moral and political order," Robespierre said, stating the mission with supreme clarity. "Half the revolution of the world is already done; the other half must be accomplished." *Sensibilité*, the "tender regard for humanity," and *bienfaisance*, the "desire to do good," became established impulses of the French Enlightenment, sparking a fever for reform that would rage eventually in London, Paris, Amsterdam, and Philadelphia. It sought change on broad humanitarian fronts such as temperance, education for women, penal reform, and an end to the slave trade, and included more explicit goals like the eradication of dueling and the building of lighthouses. "What Voltaire and his co-conspirators provided," writes Robert Darnton of the generation of the *philosophes*, "was not original matter for thought but a new spirit, the sense of participation in a secular crusade. It began with derision, as an attempt to laugh the bigots out of polite society, and it ended with the occupation of the moral high ground, as a campaign for the liberation of mankind."

For his eloquence on the subject, the Marquis de Condorcet is the individual perhaps most identified with these new possibilities. A mathematician, Condorcet became increasingly immersed in political thought in the years before the French Revolution and concluded that human goodness, like knowledge, was capable of exponential growth, leading to the perfectibility of human behavior. Just as scientific knowledge builds on itself, analyses begetting hypotheses that in turn evolve into experiments, theories, and finally solutions, so the wrongs in society might be isolated, studied, and, in a logical progression, ultimately corrected. "The sole foundation for belief in the natural sciences," he wrote, "is this idea, that the general laws directing the phenomena of the universe, known or unknown, are necessary and constant. Why should this principle be any less true for the development of the intellectual and moral faculties of man than for the other operations of nature?"

A protégé of Turgot, Condorcet had been crushed by his mentor's expulsion in 1776 from Versailles, where, as comptroller general of finances, Turgot had pushed an ambitious program of reform. "A great man," Condorcet hailed Turgot, "[who] was convinced that the truths of political and moral science are capable of the same certainty as those that form the system of physical science." Turgot, who as administrator of one of France's poorest regions (Limoges) had seen directly the results of the state's indifference, set out to redress some of the hardships endured, with mounting bitterness, by the country's lower classes. He called for an end to the *corvée*, the compulsory road-building that peasants performed for the Crown, a reassessment of property taxes, and a loosening of restrictions on agricultural trade. He initiated such useful ideas as a veterinary school for cattle owners, advocated the planting (and eating) of potatoes, and saw to it that the provinces had access to a superior rat trap newly invented in Paris. But he also called for a reform that probably hit too close to home: the curtailment of wasteful and immodest spending at Versailles. It is a pet theory of historians that had Louis XVI not given in to the demand made by the nobility to dismiss Turgot, and had Turgot's reforms been fully enacted, the French Revolution, or at least its most murderous and tragic features, might never have come to pass.

The French Revolution, in which an inept, isolated monarchy and its government met the popular anger its negligence had exacerbated, had a happier variation across the Atlantic, where the now independent Americans moved relatively smoothly from revolution to the task of creating a democratic government. The Americans, after all, had not toppled a kingdom but had simply withdrawn from one. They were not tearing something down so much as putting something up, building on the flourishing idea of colonial unity and interdependence that had been dear to Franklin and many others since the French and Indian War. "A new and intoxicating liberal idea . . . was barely beginning to enter the consciousness of American colonists," observes historian Joseph Ellis, "namely, that if all the artificial restraints and regulations imposed on human activity were removed, the result would not be chaos but harmony; moreover, that the religious and political health as well as the economic and cultural productivity of such a society would increase dramatically." The colonists' bitter experience with distant authority, the lack of representation in Parliament, unfair forms of taxation, restrictions on manufacturing, the billeting of troops, and the Crown's punishing edicts all contributed to their ideas about the new government they wished to erect.

This "seed time of the American republic" promised a glorious flowering of the Enlightenment's tenets, for the founders embraced the notion that the clarity of science could encourage new ways of thinking on moral and social grounds. As a result, the documents of national conception they created—the Declaration of Independence and the Constitution—invoked the "laws of nature," principles that guaranteed the fundamental rights of the individual in a governed society, and that were as intrinsic to humanity as Newton's maxims were to the functioning of the cosmos.

Jefferson in the Declaration attributed the document's authority to "the laws of nature and of nature's God," a phrase that recalled Newton's "laws of motion," although what differentiated Jefferson's "laws" from Newton's was that the latter's had by now been largely proven, while the "laws" Jefferson vouched for in the Declaration were more in the nature of a working hypothesis, an embryonic "science of freedom." Franklin made one small but important editorial suggestion to the efforts of his Virginia colleague. Where Jefferson referred to the principles of human equality as "sacred and undeniable," Franklin suggested, "We hold these principles to be *self-evident*," invoking a term Newton had applied to scientific truth. Like Newton's laws of motion, Franklin meant to imply, it is an observable fact that all men are created equal.

"Science and its philosophical corollaries were perhaps the most important intellectual force shaping the destiny of eighteenth-century America," Clinton Rossiter writes. "Franklin was only one of a number of forward-looking colonists who recognized the kinship of scientific methods and democratic procedure. Free inquiry, free exchange of information, optimism, self-criticism, pragmatism, objectivity—all these ingredients of the coming republic were already active in the republic of science that flourished in the eighteenth century."

To the founders, the laws of nature were not just an endearing theory of an idealized condition, but a working notion of how government must be limited to allow men to exist and thrive in their natural state, in order that they might be virtuous, creative, and happy. Perhaps no more succinct definition of America exists than to say that it is a nation in which, ideally, all persons are extended the right to pursue happiness and personal fulfillment. Previously, philosophers had spoken of man's inalienable right to "life, liberty, and *property*," which Jefferson retooled in the Declaration, replacing "property" with "the pursuit of happiness." Having diminished God's role in the scheme of things, the Enlightenment perhaps substituted happiness for a religious state of grace. After all, as Franklin said, God certainly had not made men to be *unhappy*. But the most salient point on which the founders agreed was that an individual's natural rights predate and are superior to any form of government devised by society. "The sacred rights of mankind are not to be rummaged for among old parchments or musty records," acknowledged Alexander Hamilton. "They are written, as with a sunbeam, in the whole volume of human nature, by the hand of the divinity itself, and can never be erased or obscured by mortal power."

The faith in American exceptionalism that had come ashore with the Pilgrims at Plymouth Rock had been reinforced by what struck many of the founders as the near-miraculous winning of the American Revolution, the sense that an "angel in the whirlwind" had guided the poorly armed, frequently dispirited colonists to defeat one of the foremost military powers on earth. They sensed consequently that they stood at the leading edge of a new and potentially great civilization that had been thrust forward by destiny or, as George Washington observed, "at an Epocha when the rights of mankind were better understood and more clearly defined that at any former period." Greece, Rome, the Renaissance, Elizabethan England—these glorious ages had passed: Would America's be next? "The Arts have always traveled Westward," Franklin told Philadelphia artist Charles Willson Peale in 1771, "and there is no doubt of their flourishing hereafter on our side of the Atlantic."

Franklin's certitude about the glorious American future was not simply an expression of his confidence in democratic ideals; it was based on mathematics. His *Observations Concerning the Increase of Mankind, Peopling of Countries, etc.*, written in 1751 (then revised for publication in 1755), estimated that the population of the colonies would double every twenty-five years, even without further immigration. He noted that, at that rate, within a century "the greatest number of Englishmen will be on this side of the water." The idea of the biological growth of a population—that, unchecked by war, famine, or epidemic, populations will increase exponentially—was not addressed solely by Franklin in this era; Robert Wallace took up the issue in 1753 in *A Dissertation on the Numbers of Mankind in Ancient and Modern Times*. But considering that Franklin's writing on population was done by guesswork (he died before the completion of the first United States census in 1790), his estimates have proved uncannily accurate. His prediction, made in 1755, that within a hundred years more English-speaking people would reside on the American side of the Atlantic, came true a century later, around 1851; his estimate that America's population would double every twenty-five years also proved correct: Applied to the 1790 census, which found a population of 3,929,214, this implied an 1890 population of 62,867,424. The actual 1890 census showed a population of 62,947,714, meaning Franklin, predicting 135 years into the future, was off by only 0.13%.

Numerical estimations led Franklin to his promising conclusions, but he had also come simply to possess tremendous faith in Americans as individuals and in the collective American future. As his *Autobiography* would make clear, Franklin saw his own life story—that of a man who had emerged from the obscurity of a Boston candle-maker's shop to become a leading publisher, diplomat, and science celebrity—as one that could serve as an example to others; and he believed it to be far from unique, for it was paralleled by the trajectories of many men he admired—Jefferson, Adams, Washington, Rittenhouse, Peale, John Bartram, Colden, and Kinnersley, among others. What he valued particularly were the American traits of skepticism and independence of thought, and the willingness to judge personal worth on criteria other than social status; and he was heartened by what he perceived to be a growing thirst among his countrymen for books, knowledge, and conversation. "Instinctively more comfortable with democracy than were his fellow founders," suggests biographer Walter Isaacson, "[Franklin] had faith . . . that a new nation would draw its strength from what he called 'the middling people.'"

As for the further settlement of the country, Franklin turned again to the image of the polyp, the tiny self-procreating animal that had long intrigued him. Once settlements grew and prospered, some of their residents would tear away and plant themselves elsewhere, and in another generation a new community would grow; and this process would be endlessly repeated. "Take away a Limb, its place is soon supply'd; cut it in two, and each deficient Part shall speedily grow out of the Part remaining. Thus if you have Room and Subsistence enough, as you may by dividing, make ten Polypes out of one, you may of one make ten Nations, equally populous and powerful; or rather, increase a Nation ten fold in Numbers and Strength."

The Constitution of the United States was to be—along with the lightning rod, Newton's *Principia*, and the music of Handel and Mozart—one of the Enlightenment's supreme accomplishments. The delegates to the Constitutional Convention who gathered in Philadelphia in May 1787 were fully aware of how important their efforts would be, and how devastating would be any failure. "It is more than probable that we are now digesting a plan which in its operation will decide forever the fate of a republican form of government," observed James Madison of Virginia. "If we do not give to that form due stability and wisdom," added Hamilton of New York, "it will be disgraced among ourselves, disgraced and lost to mankind forever."

The pressures on the delegates were great. The Declaration of Independence of 1776, forged in war, was based on a simple idea that had served an embryonic republic of colonies united against a common enemy. The Articles of Confederation, the document that provided the basis of a centralized government from 1781 until 1789, were already proving inadequate to run a functioning nation. Among its other shortcomings, the Articles failed to create an infrastructure for national government: Congress was given no power to coin money, levy taxes, or regulate interstate commerce; there was no judicial system and no chief executive, or president. Just that March, a former army captain named Daniel Shays had led two thousand farmers frustrated with the Massachusetts state legislature's neglect of their property rights in an armed assault on the arsenal at Springfield. Already the phrase "not worth a Continental" mocked the government's near-insolvency, and it was said of the former colonies, "Thirteen staves and never a hoop will not make a barrel!" In the absence of sufficient action from the men in Philadelphia, the fledgling United States might simply collapse from within.

Of all the delegates, Washington was closest to Franklin in experience and past acquaintance: Both had taken part in British military efforts against the French and Indians (Franklin in a supporting role), served together as members of the Continental Congress, and been central to the success of the Revolutionary War. Although Washington was every inch the Virginia gentleman and Franklin the cultured Philadelphian, they were—along with Hamilton, a New York delegate originally from the West Indies— perhaps the least attached to sectional interests; they had become accustomed to thinking of the emerging United States as a whole. Franklin was also unusual among the delegates in that he had been away from America during its recent formative struggle and had worn no uniform nor led soldiers in battle. At age eighty-one, he was, as historians like to note, more a founding *grandfather*, older by at least a generation than his colleagues, and was also in failing health. Many of his speeches had to be read for him by James Wilson, his fellow delegate from Pennsylvania, although James Madison later recollected that Franklin did on occasion "make short extemporaneous speeches with great pertinence and effect."

Anticipating great divisiveness on the many issues awaiting deliberation, Franklin urged that "the wisest must agree to some unreasonable things so that reasonableness of more consequence may be obtained." He encouraged the delegates to devote their energies, according to Rossiter, "not in associating with their own party and devising new arguments to fortify themselves in their old opinions but [to] mix with members of opposite sentiments." Franklin believed wholeheartedly that, "By the Collision of different Sentiments, Sparks of Truth are struck out, and Political Light is obtained," although if democracy was to work, it would rely, as Rossiter notes, on "men with a nice feeling for the proper balance between faith and skepticism, principle and compromise, tenacity and reconciliation"—men, in other words, like Benjamin Franklin.

"The convention did not need additional gladiators," historian William Carr writes. "It needed a delegate who . . . understood that group action is, almost always, a result of compromise."

There was, however, to be no compromise on one of the Convention's first pieces of business. Franklin had nominated his grandson William Temple Franklin—"Franklinet" of the Paris days—for the job of convention secretary, but Temple's own reputation for seriousness was not strong, and his father, Franklin's son, William, the former royal governor of New Jersey, had never repudiated his loyalist ties and was now in exile. Another of Franklin's ideas not accepted was that for a unicameral legislature, which was then used in Pennsylvania. Like many of the large-state delegates, he favored a single legislature in which states would be represented proportionate to their populations. The smaller states, like Delaware and Connecticut, naturally opposed such a plan, preferring equal representation in a bicameral system. Franklin feared that two congressional bodies would tend to hopelessly nullify one another and entangle the government's affairs, like a snake with two heads. "She was going to a Brook to drink," he told the Convention, inventing a fable for the purpose of his argument, "and in her Way was to pass thro' a Hedge, a Twig of which opposed her direct Course; one Head chose to go on the right side of the Twig, the other on the left; so that time was spent in the Contest, and, before the Decision was completed, the poor Snake died with thirst."

What perhaps came as most unexpected was Franklin's suggestion, when the Convention seemed hopelessly deadlocked over the representation issue, that each session open with prayer. "I have lived, Sir, a long time," he declared, addressing Washington, the Convention's president, "and the longer I live, the more convincing proofs I see of this Truth, that God governs in the Affairs of Men. And if a Sparrow cannot fall to the Ground without his notice, is it probable that an Empire can rise without his Aid?"

Had Franklin's religious outlook looped back around to the Congregationalist faith of his father? Or did he simply want to remind the delegates of the importance of their work? Perhaps he was concerned that the failure to resolve the question of legislative representation might cause the entire enterprise to founder. "When the country's oldest Deist issued an appeal for religious intervention," notes historian Carol Berkin, "it was obvious the convention had entered its darkest hour."

Like many of his countrymen, Franklin had been delightfully surprised by the success of a war that had so often seemed unwinnable, and as an honest skeptic he was forced to ponder if indeed "the Father of Lights" was not now illuminating America's way. His Deistic perspective, after all, always allowed for divine action in special circumstances. "If God governs, as I firmly believe," Franklin had written of Britain's efforts to suppress the Revolution, "it is impossible such Wickedness should long prosper." Now, with the separation complete and the task of establishing a new government at hand, he insisted: "I can hardly conceive [that] a Transaction of such momentous Importance to the Welfare of Millions now existing, and to exist in the Posterity of a great Nation, should be suffered to pass without being in some degree influenc'd, guided, and governed by that omnipotent, omnipresent, and beneficent Ruler, in whom all inferior Spirits live, and move, and have their Being." Thus, the man whose lightning rod had challenged the notions of faith and divinity in the revolutionary age could, in the end, not resist praising the deity for seeing the revolution safely through.

Out of respect for its author, Franklin's motion was briefly considered, then discarded. Yet, it may well have helped ease the body's contentious mood, reminding the delegates of the solemn enterprise in which they were engaged. Indeed, Franklin was one of several delegates appointed to a "committee of compromise" asked to recommend a solution to the representational impasse. Despite his strong feelings about the issue, he gradually accepted the fears of the smaller states that they would be overwhelmed by larger states with a greater number of representatives, and accepted a compromise put forward originally by Roger Sherman of Connecticut: The United States would have a bicameral Congress, with each state to have equal representation in the Senate, and a House of Representatives in which the number of members was to be based proportionately on state population. In each body, the senators and representatives were free to vote as individuals, not merely as part of their state's delegation, and the approval of both houses of Congress would be required for a bill to become law. As a guard against the potential excesses of the Senators, who, since they would be elected by their state legislatures, were expected to be prominent or wealthy men, it was established that appropriations bills would have to originate in the House. The Southern states were handed the right, as they had enjoyed under the Articles of Confederation, to count each slave as three-fifths of an individual for purposes of determining the number of representatives each state could send to Congress.

In addition to his work on the compromise, Franklin helped establish the federal judiciary, advocated a shorter waiting time for the naturalization of new citizens, spoke out for measures to protect dissent, and swung support against an absolute veto power in the presidency.

He had another important role. As the Convention was held in Philadelphia, Franklin was something of its unofficial host, and his home on Market Street, where he had ordered his dining room enlarged for the purpose, became an informal gathering place for the delegates, many of whom were staying at the nearby Indian Queen. Overseen now by Franklin's daughter, Sally (his wife, Deborah, had died in 1774), the Market Street household centered in summer months around an inner courtyard, where Franklin liked to receive guests while sitting beneath a shady mulberry tree, likely a token of his enduring interest in silk culture. (Franklin and his friend Ezra Stiles had long been advocates of an effort to make America "a nation of silk growers," but a depression in the silk-weaving industry in England in the 1760s and the emergence of the use of cotton, which was cheaper to raise and more popular, helped to doom the prospects for silk cultivation in the colonies.)

Having spent so much time away from it, Franklin appeared to savor life in his own home, with his garden, his books and scientific equipment, his doting daughter and grandchildren, and a constant flow of interesting visitors. George Washington came by to examine a new "washing mangle" that someone had offered Franklin as a novel method of cleaning clothes, Noah Webster to discuss his and Franklin's mutual interest in language and phonetics; and John Fitch, inventor of an early steamboat, sought Franklin's approval for his work. Some of the delegates took a field trip to John Bartram's garden, where they examined a tree named for Franklin, the *Franklinia alatamaha*. But Franklin himself remained the chief attraction.

"There was no curiosity in Philadelphia which I felt so anxious to see as this great man," wrote Manasseh Cutler, a visiting clergyman and botanist from Massachusetts. Certain that such a celebrity would be unapproachable, Cutler was taken aback when he encountered Elbridge Gerry, a Massachusetts delegate, in the street, and Gerry, saying he was on his way to Franklin's house, invited him to tag along. "I hesitated; my knees smote together," Cutler confessed.

They found Franklin sitting beneath the mulberry tree entertaining a group from the Convention. Cutler saw "a short, fat, trunched old man, in a plain Quaker dress, bald pate, and short white locks, sitting without his hat," who spoke in a quiet voice but with a "countenance open, frank, and pleasing." Franklin appeared elated to welcome a man of science, and insisted on showing Cutler a number of curiosities, including a machine purchased years before that replicated the actions of the human heart and circulatory system; a two-headed snake preserved in a jar (no doubt the inspiration for the parable he had told in the Convention); and a huge illustrated edition of Linnaeus's *Systema Vegetabilium*. Cutler was charmed when Franklin, with an old man's pride, refused any assistance lifting the heavy Linnaeus volume from a lower shelf.

"I was highly delighted with the extensive knowledge he appeared to have of every subject, the brightness of his memory, and clearness and vivacity of all his mental faculties," Cutler recorded. "Notwithstanding his age, his manners are perfectly easy, and every thing about him seems to diffuse an unrestrained freedom and happiness. He has an incessant vein of humor, accompanied with an uncommon vivacity, which seems as natural and involuntary as his breathing." Franklin also demonstrated two products of his own recent ingenuity—a rocking chair in which a foot pedal operated a fan that kept flies away from the sitter, and a wooden device he called "The Long Arm," devised for grasping books from the upper shelves of his library. He confided to his visitor, "Old men find it inconvenient to mount a ladder or steps for taking down books from high shelves, their heads being sometimes subject to giddiness." At one point Franklin began telling Cutler an amusing anecdote from that day's discussion at the Convention, but one of the other delegates immediately cut him off, reminding him of the rule that its deliberations were not to be spoken of in public until the Convention adjourned.

Franklin, who once termed the century in which he lived "The Age of Experiments," ultimately gave the experiment that was the Constitution of the United States of America his blessing, claiming to be pleased especially with the provision that it remain open to amendment. "Experience keeps a dear school," *Poor Richard's* had observed, and Franklin assumed that even well-crafted plans of governance, once put into effect, could not go long without adjustment. "I confess that I do not entirely approve this Constitution at present," he said in his last remarks, "but Sir, I am not sure I shall never approve it: For having lived long, I have experienced many instances of being obliged by better information or fuller consideration to change opinions even on important subjects which I once thought right but found to be otherwise." He warned that any government would be in danger if its people did not tend to it, and that America's government, like others before it, "is likely to be well administered for a course of years, and can only end in despotism . . . when the people shall become so corrupted as to need despotic government, being incapable of any other." But he would support the Constitution's ratification "because I expect no better, and because I am not sure that it is not the best."

A small crowd had gathered outside the State House door on the rumor that the Convention had completed its work. As the delegates emerged, Elizabeth Willing Powell, a prominent local intellectual and wife of Samuel Powell, the mayor of Philadelphia, caught a glimpse of Franklin's unmistakable profile. "Doctor, what have we got," she asked, "a republic or a monarchy?" Franklin, finally free to speak of the Convention's achievement, called out, "A republic, if you can keep it."

With the signing of the Constitution, the first chapter in the history of the United States, and the last in Franklin's long public life, appeared to have come to an end. His snatching of thunder from God had given symbolism to the American Revolution, his diplomacy had won decisive support abroad, and his appreciation of the value of experiments and his faith in compromise had guided the creation of the document that completed the work of the founders. But, as he had once confided to his friend George Whitefield, "Life, like a dramatic Piece, should not only be conducted with Regularity, but methinks it should finish handsomely. Being now in the last Act, I begin to cast about for something fit to end with." In fall 1787 Franklin found that something: He decided to test the fragile experiment in democracy he'd done so much to create—to throw a few lightning bolts of his own—by becoming the first prominent American to call for the abolition of slavery.

It must have been apparent to all the founders that the slavery compromise in the Constitution—"The Migration or Importation of Such Persons as any of the States now existing shall think proper to admit, shall not be prohibited by the Congress prior to the Year one thousand eight hundred and eight"—was a feeble half-measure, meant to win the inclusion of the Southern states in the Union and postpone any serious reckoning with the issue. So uncomfortable were the delegates with the subject, that nowhere in the Constitution appeared the words "slavery," "slaves," or even "Negroes." Franklin, having stressed at the Convention the necessity of compromise, nonetheless knew that the omission was potentially costly, and he recognized as misguided the faith held by some of his colleagues that the "peculiar institution" of slavery might soon fade away of its own absurdity. As a British subject, writes biographer H. W. Brands, "Franklin had been able to countenance slavery as one public vice among many received from the past. As an American, he could no longer countenance it, for the new nation could not abide public vice—certainly not of the magnitude of slavery—without jeopardizing its very existence."

If Franklin's eight decades had given him anything, it was a healthy regard for the value of anticipating political crises before they grew intractable. In 1754, his Plan of Union for the colonies was adopted by delegates at a gathering in Albany, but was never approved by the colonies themselves. "On Reflection it now seems probable," he wrote in 1789, "that if the [Albany] Plan or some thing like it, had been adopted and carried into Execution, the subsequent Separation of the Colonies from the Mother Country might not so soon have happened, nor the Mischiefs suffered on both

sides have occurred, perhaps during another Century." Because the British cited the expense of defending the colonies during the French and Indian War as the reason for enacting various taxation schemes, such as the Stamp Act, that so infuriated the colonists, Franklin couldn't help wondering if a unified group of colonies might have been able to nullify this need and its related demand. "But such Mistakes are not new," Franklin mused. "History is full of the Errors of States and Princes."

A number of leading Quakers, who had shown themselves on both sides of the Atlantic to be a kind of conscience to the eighteenth century, convinced Franklin shortly after his return to Philadelphia from France to become head of the "Pennsylvania Society for Promoting the Abolition of Slavery and for the Relief of Free Negroes Unlawfully Held in Bondage," the first abolitionist group in the New World. In an unmistakable allusion to the delegates of the Constitutional Convention, the Society soon called on "those persons, who profess to maintain for themselves the rights of human nature, and who acknowledge the obligations of Christianity, to use such means as are in their power, to extend the blessings of freedom to every part of the human race."

What made Franklin's stand against slavery such a fitting culmination to his long life was that his views on the subject had matured over a great many years, particularly during his time abroad in England and France. There he had been exposed directly to the abolitionists and *philosophes* who were among the first to condemn slavery's evil. As early as 1721, Montesquieu pointed out that slavery was so immoral it corrupted the values of both slaves and masters and that, despite appearances, it did not represent sound economics. By the 1770s Franklin's acquaintance the abbé Guillaume Raynal was warning of the despotic excesses of colonialism. Calling the meeting of the European and native worlds "one of the most important events in the history of the human species," Raynal moralized heartily against slavery: "Nothing . . . is more miserable than the condition of a black, throughout the whole American Archipelago . . . condemned to a perpetual drudgery in a burning climate, constantly under the rod of an unfeeling master." The Marquis de Condorcet, who along with Lafayette in 1788 founded the French antislavery group "Les Amis des Noirs," was the author of *Reflexions sur l'Esclavage des Nègres* (1782), another essential antislavery text. While the *philosophes* gave the antislavery cause its initial intellectual heft, it would first assume the form of a crusade in England, guided by Evangelicals William Wilberforce and Granville Sharp and egged on by the Philadelphia Quaker Anthony Benezet, who in 1772 urged Franklin, then in London, to "let, not only the deep sufferings and vast destruction of these our deeply distressed fellow Men, but also the corrupting effects it has on the hearts of their lordly oppressors, be, as the scripture expresses it, precious in thy eyes."

Like most eighteenth-century Americans, Franklin had grown up in a culture of legalized bondage—of slavery, apprenticeship, and indentured servitude—and like his father before him, he had occasionally sold and traded in slaves. His *Pennsylvania Gazette* regularly printed, along with notices about runaway apprentices and servants, advertisements offering "Likely Negroes" for sale, although Franklin also published Quaker abolitionist tracts such as *A Brief Examination of the Practice of the Times* by Ralph Sandiford (1729) and Benjamin Lay's *All Slave-Keepers That Keep the Innocent in Bondage* (1737). Franklin's own experience at the hands of his brother James, the "harsh and tyrannical treatment" that gave him his "aversion to arbitrary power," no doubt predisposed Franklin to sympathize with those unjustly subjugated or abused, and he more than once made public his opposition to whites' cruelty toward the Indians of Pennsylvania as well as to the British practice of impressing seamen. But as a propagandist, he was not above evoking racial anxieties. In "Plain Truth," his 1747 pamphlet intended to call attention to the danger of attack from privateers, who, it was rumored, had been joined by runaway slaves, Franklin alerted his neighbors to "the utmost Horror" of white "Wives and Daughters . . . subject to the wanton and unbridled Rage, Rapine and Lust of *Negroes, Molattoes*, and others, the vilest and most abandoned of Mankind." In his 1755 treatise on population, he questioned whether owning slaves was economically feasible in the long term, since the slave had no motivation to work, required close supervision, needed to be clothed, fed, and cared for, and was "by Nature a Thief"—a phrase he softened in a later edition to suggest blacks were perhaps driven to steal and misbehave "from the nature of slavery." Ever a believer in an honest day's work, Franklin was also alert to the desultory effects of slavery on the slave-owning class itself. "The Whites who have Slaves, not labouring, are enfeebled," he commented, "the white Children become proud, disgusted with Labour, and being educated in Idleness, are rendered unfit to get a Living by Industry."

The sea change in his thinking was first noticeable during his initial mission to England, when he became acquainted with "Dr.

Bray's Associates for Founding Clerical Libraries and Supporting Negro Schools." Thomas Bray was an Anglican minister devoted to establishing libraries both in England and in the colonies, and a posting in Maryland had convinced him of the need for schools for black children. When a Bray school opened in Philadelphia in 1758, Deborah Franklin went to visit—probably at her husband's urging—and was so impressed at hearing the children recite, she immediately enrolled her own slave, Othello. Franklin's slaves Peter and King, who had accompanied him from Philadelphia to London, seem to have taken the best advantage of their master's enlightened attitude. King ran away to reside with an English lady who taught him to read, write, and play the violin; Peter's stubbornness and numerous faults Franklin chose simply to ignore, noting that in this way they continued to "rub on pretty comfortably."

Since a frequent charge against black people was that they were "deficient in natural understanding," Franklin's empirical mind was drawn to the possibility of testing the allegation. In a 1763 letter to John Waring, secretary of Dr. Bray's group, Franklin reported that having visited one of the schools operating in Philadelphia, he was struck by the children's immense capacity for learning, and admitted that his earlier beliefs required revision. "I was on the whole much pleas'd, and from what I then saw, have conceiv'd a higher Opinion of the natural Capacities of the black Race, than I had ever before entertained. Their Apprehension seems as quick, their Memory as strong, and their Docility in every Respect equal to that of white Children. You will wonder perhaps that I should ever doubt it, and I will not undertake to justify all my Prejudices, nor to account for them." To Condorcet, a decade later, Franklin expressed the view that blacks seemed to him "not deficient in natural Understanding, but they have not the Advantage of Education." He had glimpsed a core truth: Human beings' ability to aspire and learn was due not only to their innate qualities but also to the conditions in which they were made to exist.

Thus Franklin became an abolitionist in stages—first recognizing that blacks would thrive if educated, second that they are equal and not innately inferior to whites, and finally that slavery itself was a monstrous injustice and a stain on the emerging American character. In 1772 he wrote to the *London Chronicle* complaining of the pretense of the Sommersett case, in which a slave by that name arriving in England was ordered freed:

> Can sweetening our tea, &c. with sugar, be a circumstance of such absolute necessity? Can the petty pleasure thence arising to the taste, compensate for so much misery produced among our fellow creatures, and such a constant butchery of the human species by this pestilential detestable traffic in the bodies and souls of men? *Pharisaical Britain!* to pride thyself in setting free *a single slave* that happens to land on thy coasts, while thy merchants in all thy ports are encouraged by thy laws to continue a commerce whereby so many *hundreds of thousands* are dragged into slavery that can scarce be said to end with their lives, since it is entailed on their posterity!

The following year he had a sympathetic visit in London with Phyllis Wheatley, the gifted twenty-year-old black poet who had accompanied her master, Nathaniel Wheatley, from Boston. There may have been some fear on the part of Master Wheatley that Franklin intended, in the wake of the Sommersett case, to use his influence to assist Wheatley in declaring her status as a free woman, for in a subsequent letter Franklin told his cousin Jonathan Williams that "I understood her Master was there and had sent her to me but did not come into the Room himself, and I thought was not pleased with the visit."

While the Constitution expressed concern over the importation of slaves, it hardly addressed the status of the growing number of slaves already in America. Even with a halt to the transatlantic slave trade, slavery in America would only fester and grow, "a cancer that we must get rid of," as Charles Thomson, secretary of Congress, characterized it to Thomas Jefferson. Jefferson had written against slavery as a moral and economic erratum in his *Notes on Virginia*, and denounced importation in an early draft of the Declaration of Independence. In 1789, Jefferson proposed a moratorium on slavery in new territories opening in the West, an early peek at the rhetorical battleground where much debate would take place in the next century, and—converted as Franklin had been by the evidence that blacks could be educated—sent to Condorcet some impressive computations by the American Negro mathematician Benjamin Banneker.

The positions on either side of a debate with tragic implications for America were already becoming entrenched—abolitionists, holding firm that slavery could not be tolerated and that it should be eradicated sooner rather than later, and representatives of the Southern planters, insisting they would not give up their economy and way of life nor live with an emancipated population of ex-slaves. Southerners also questioned the national government's authority to regulate what many saw as an issue of property rights to be handled by the individual states. Logistical issues lent an intractable quality to the situation: There were 694,280 slaves registered by the census of 1790, all but about 45,000 of whom lived in the South. Few people, north or south, were inclined to view the future America as a biracial society, and the costs of resettling America's slaves elsewhere, even if physically possible, would be staggering.

Franklin nonetheless saw the harm in postponing a reckoning. In November 1789, after the Constitution had been ratified, he signed, and probably wrote, an address on the topic to the American public that demanded an end to slavery's "atrocious debasement of human nature," as well as an attached program that prescribed educating black children and employing newly freed Negroes. The Pennsylvania abolition group thus tried to quell one common objection to emancipation by suggesting that they would also work to ease the blacks' adjustment to freedom.

In early February 1790 the Quakers petitioned Congress, then meeting in New York, for an immediate end to the slave trade, followed the next day by an antislavery memorial from the Pennsylvania Abolition Society, signed by Franklin. "They have observed with great Satisfaction that many important and salutary powers are vested in you for 'promoting the welfare and securing the blessings of liberty to the People of the United States,' " Franklin told Congress of his abolitionist colleagues, "and as they conceive that these blessings ought rightly to be administered, without distinction of Color, to all descriptions of People, so they indulge themselves in the pleasing expectation that nothing which can be done for the relief of the unhappy objects of their care will be either omitted or delayed." Franklin urged Congress to "devise means for removing this Inconsistency from the Character of the American People."

Southern representatives reacted with alarm. They wanted to believe that the twenty-year protection on the slave trade meant also a prohibition on mentioning the subject, and they read in Franklin's words the outlines of an inevitable crusade for emancipation, for he not only had reminded them that slavery and the slave trade were inconsistent with the values of the young country, but challenged the idea that Congress was forbidden to act on the issue. Emancipation, thanks in large part to Quaker advocacy, could even be said to be in the air; Vermont, New Hampshire, Pennsylvania, Massachusetts, Connecticut, Rhode Island, and New York had all recently enacted emancipation or gradual-abolition laws; New Jersey had banned the importation of slaves; and even Virginia had passed a law allowing owners to manumit slaves, which by 1790 had created a population there of twelve thousand free Negroes.

The near-simultaneous arrival on February 11 and 12 of these petitions—two from Quakers in the mid-Atlantic states and one from the Pennsylvania Abolition Society bearing Franklin's signature—guaranteed they would be taken up, and because the deliberations of Congress were public, the result was the first truly national discussion of the issue. It was a dialogue that quickly became rancorous, with Quakers and other abolitionists packing the gallery of Congress, as one Southerner complained, like "evil spirits hovering over our heads."

The emotional debate that ensued rehearsed the major points of contention that would characterize the abolitionist cause in the nineteenth century, including claims by Southerners that the Constitution, by its very evasiveness and refusal to address the subject, had assented to allow the states alone to decide whether or not to maintain slavery; that slavery was authorized by Scripture; that slaves were necessary for the Southern economy ("It is well known the rice cannot be brought to market without these people"); and that emancipation would mongrelize the white race. The representatives of the Southern states expressed open contempt for the Quakers, reminding their colleagues that they had sat out the Revolutionary War on account of their "principles," and depicted them as frivolous dreamers and eccentrics. A Quaker from Maryland named Warner Mifflin soon obliged these characterizations by assuring Congress that his resolve against slavery had come to him upon being struck by lightning.

Georgia's James Jackson charged that the Quakers were distributing "dirty pamphlets representing Negroes" and challenged them to come up with a means of compensating slaveholders for the losses they were asking them to incur. "Is the whole morality of the United States confined to the Quakers?" he asked. "Are they the only people whose feelings are to be consulted on this occasion? Is it to them we owe our present happiness? Was it they who formed the Constitution? Did they, by their arms, or contributions, establish our independence? Why then, on their application, shall we injure men, who, at the risk of their lives and fortunes, secured to the community their liberty and property?"

A few weeks later, an ailing Franklin made what would be his last public effort, publishing under the name "Historicus" a faux-arabesque tale that mocked the proslavery speeches heard in Congress. Such pseudo-oriental fables were a popular form of satire in the eighteenth century, and Franklin had dabbled in them before. "Historicus" wrote that he had found some parallels between a speech given by Representative Jackson and one written a hundred years ago by an Algerian pirate, Sidi Mehemet Ibrahim. The pirate explained why it was necessary to enslave Christians, and worried about what Christians might do if set free, where they would go, or if they might pillage their former masters' property or attempt to intermarry, and questioned that the government could adequately compensate their masters. Wryly suggesting that Jackson could not possibly have copied the pirate's speech, Franklin nonetheless noted that the words were similar. "If . . . some of its reasonings are to be found in his eloquent speech, it may only show that men's interests and intellects operate and are operated on with surprising similarity in all countries and climates." The whole thing was vintage Franklin, the tale's anonymity and its sharp satire offering a kind of bookend to his long-ago literary debut when, as the widow Silence Dogood, he slid his judicious prose beneath the door of his brother's *New-England Courant.*

Unfortunately, this time the man who had snatched thunder from God and the scepter from kings was up against perhaps an even more powerful force, and his petition against the sinful paradox of slavery in the democratic United States, though hotly discussed, was in the end cast aside, unheeded. "Do these men expect a general emancipation of slaves by law?" demanded Thomas Tudor Tucker of South Carolina. Referring to Franklin as "a man who ought to have known the Constitution better," he warned prophetically, "This would never be submitted to by the Southern states without a civil war." Jackson of Georgia echoed the theme: "The people of the southern states will resist one tyranny as soon as another; the other parts of the continent may bear them down by force of arms, but they will never suffer themselves to be divested of their property without a struggle." In shunting the calls for abolition aside, the Southerners debuted a tactic they would use frequently in the years ahead. Filibustering for two successive days, Jackson and his cohorts gave voice to every known proslavery argument, and when those were exhausted improvised new ones, declaring that slavery was a gift to the blacks of Africa, who otherwise would remain in squalor, and that abolitionists were simply religious fanatics with guilty consciences. To the chagrin of those watching from the gallery, Northern representatives eventually gave up their counterarguments.

Vice President John Adams would, a few months later, confide to a friend that those members of Congress who took seriously the "self-constituted" abolition societies and "the silly petition of Franklin and his Quakers" were allowing themselves to be distracted from more important business facing Congress, such as establishing national economic policy and financial stability. Many others concurred that the time was not right to be meddling in so difficult an area, and that antislavery arguments were, for now, "a matter of moonshine." But it wasn't only that Congress's agenda happened to be full. Simple racial fears and antagonisms were also at play. Antislavery Northerners were no doubt themselves unsure how far they were willing to go on behalf of a people they believed to be inferior and whose post-emancipation fate they could only dimly imagine at best.

One Southern representative, William Loughton Smith of South Carolina, in questioning Franklin's motives for forcing so unwieldy a subject, was so bold as to assure Congress that Franklin's thinking had likely become addled with old age, noting "even great men have their senile moments." Pennsylvania's delegates had failed to talk down the emotional Southern faction, but so personal an insult to their state's greatest light could not be tolerated. Benjamin Franklin's views, the irreverent Smith was assured, demonstrated the "qualities of his soul, as well as those of his mind," and showed that only he could still "speak the language of America, and . . . call us back to our first principles."

In the spring of 1790, it was Franklin himself who was being called back. Suffering from gout, kidney stones, a recurring respiratory ailment, and a sprained wrist that had never fully healed from a fall he had taken two years earlier on the steps of his garden, he eased the pain by sitting and reading for hours in a special slipper-shaped bathtub, and with small doses of laudanum. He took his decline stoically. He told Dr. John Jones, his physician, that God had given him so many blessings in life, raising him "from small and low beginnings to such high rank and consideration among men" that he had "no doubt but his present afflictions were kindly intended to wean him from a world in which he was no longer fit to act the part assigned him." His once robust appetite diminished, he began to lose weight and, recognizing that the end was near, exchanged farewells with some of his oldest correspondents. Ezra Stiles assured him, "You have merited and received all the Honors of the Republic of Letters, and are going to a World where all sublunary Glories will be lost in the Glories of Immortality."

Franklin wrote to the man who alone among other Americans

was his equal in public stature. Operating thousands of miles apart, one at Passy, the other at the head of the Continental Army, they had together helped conduct a nation into being. "For my own personal Ease," he confided to George Washington, "I should have died two Years ago, but tho' those Years have been spent in excruciating Pain, I am pleas'd that I have liv'd them, since they have brought me to see our present Situation. I am now finishing my 84th, and probably with it my Career in this Life; but in whatever State of Existence I am plac'd hereafter, if I retain any Memory of what has pass'd here, I shall with it retain the Esteem, Respect, and Affection with which I have long been, my dear Friend, Yours most sincerely, B. Franklin."

A reply came swiftly:

If to be venerated for benevolence, if to be admired for talents, if to be esteemed for patriotism, if to be beloved for philanthropy can gratify the human mind, you must have the pleasing consolation to know that you have not lived in vain; and I flatter myself that it will not be ranked among the least grateful occurrences of your life to be assured that so long as I retain my Memory, you will be thought on with respect, veneration and affection by, Dear Sir, Your sincere friend Obedient Humble Servant,

G. Washington.

Jefferson, back from his ministerial posting in Paris and en route to Congress in New York, stopped briefly at Market Street for a last visit. He and Franklin mulled over the current state of affairs in France and discussed mutual friends there. The assault on the Bastille had taken place the previous July, and both the ancien régime and the church had been by now greatly compromised. Franklin would surely have mentioned a letter received recently from Lavoisier, in which the chemist explained that there appeared to be no turning back the Revolution. Moderates like himself, Lavoisier said, were concerned "that circumstances have carried us too far, that it is unfortunate to have been obliged to arm the common people," for "it is ill-advised to place force in the hands of those who should obey." To Franklin he lamented the fact that "you are so far from France; you would have been our guide, and you would have marked out for us the bounds that we should not have exceeded." At this early stage of the French Revolution, it still seemed faintly possible that a liberal regime like the one recently founded by Franklin and the Americans could emerge in France, and Franklin jotted optimistically in his letters to Paris his satisfaction that both nations were now seeking to establish better governments. But among their other differences from Americans, the French lacked the latter's long experience with self-government, what André Maurois called "the solid political education which had been imparted by the town meeting," and as Lavoisier's colleague the Comte de Mirabeau had darkly warned, "When you undertake to run a revolution, the difficulty is not to make it go; it is to hold it in check."

Franklin would not live to know of the worst ravages of the Terror that would consume Lavoisier, Condorcet, Bailly, Louis XVI, Marie Antoinette, and so many more, including that proud defender of lightning rods, Robespierre, and utterly sweep away the society in which he had so recently basked. It was Lavoisier and Bailly who had joined Franklin in warning the Mesmer committee of the peril of mob psychology. One of the most tragic losses was Condorcet, whose devotion to liberal rationalism ultimately proved to be as much at odds with the authority of the revolution as it had with that of the monarchy. Condemned to death by his revolutionary brethren in October 1793, he tried to disappear but was arrested at an inn, the story goes, after giving away his aristocratic background by revealing he had not the slightest idea how many eggs are needed to make an omelet.

The murder of the brilliant Lavoisier had seemed particularly needless. "It took them only an instant to cut off that head, and a hundred years may not produce another one like it," one of his colleagues remarked. Another of Franklin's old transatlantic associates to receive rough treatment was Joseph Priestley, who in addition to being a first-rate natural philosopher was also an outspoken Unitarian minister. Having enraged conservative Britons with his criticisms of the Bible and his recent support for the revolution in France, his home was sacked on July 14, 1791, the second anniversary of Bastille Day, a mob destroying his manuscripts, books, and furniture, and using his electrical apparatus to try to set the building on fire. Priestley eventually emigrated to America, settling in the town of Northumberland, Pennsylvania, where he died in 1804.

One of the final glories of monarchial France had been the 1788 funeral of Buffon, the "French Pliny," who had done so much to introduce Franklin's science in France. His memorial procession, a massive spectacle watched by twenty thousand Parisians, surpassed even the honor paid to Voltaire a decade before. Sixty priests, more then thirty-five groups of schoolchildren, and hundreds of dignitaries walked in tribute alongside columns of servants and costumed horses. But when the revolution arrived, Buffon's beloved royal garden, now renamed the Jardin des Plantes, was marked for destruction. It was saved at the last moment by quick-thinking friends who reminded the mob that the garden grew herbs for curing sickness, and that the adjacent museum's chemistry equipment might be adapted for making gunpowder. Buffon's only son, a scapegrace known by the nickname *Buffonet*, went to the guillotine, shouting defiantly from the tumbrel, "My name is Buffon!"

Franklin, having lived for nine years in France, understood perhaps better than any American the official mismanagement, corruption, dissolution, and lack of good judgment that had helped embolden the revolution, but he also could not forget how necessary the actions of the now discredited regime had been in securing America's independence. He may well have emphasized to Jefferson what he had earlier told an English friend, John Baynes: "I trust we shall never forget our obligations to France, or prove ungrateful."

One discovery Franklin had been pleased to make upon his return from France to America was that, in his absence, his house had been struck by lightning. Initiating a renovation of his Market Street house, he had located an original 1752 lightning rod still intact on the roof, its copper tip thoroughly bent and blackened. As there was no other damage to the house visible, Franklin concluded happily that "at length the invention has been of some use to the inventor." Of course, he had never lost faith in the fundamental design of the rod he had described in the *Poor Richard's* of 1753. "Indeed, in the construction of an instrument so new," he told Ebenezer Kinnersley, "and of which we could have so little experience, it is rather lucky that we should at first be so near the truth as we seem to be, and commit so few errors."

EPILOGUE

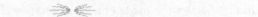

BENJAMIN FRANKLIN'S REFUSAL TO PATENT HIS "INSTRUMENT SO NEW" likely contributed to the competitive free-for-all that began to characterize lightning rod design, manufacture, and sales within a few decades of his death. In Franklin's day lightning rods had been used chiefly for large public structures and by the privileged, who alone possessed large manor houses and other valuable properties. Kinnersley, probably America's first lightning rod salesman, promoted heavily to the elites. What neither Franklin nor Kinnersley could fully know was that, by the 1840s, a revolution would occur in how goods were manufactured. Whereas in the late colonial era a lightning rod might be made to order by a local blacksmith or mechanic, by the 1840s there were at least fifteen lightning rod factories in America, and the devices were being marketed directly to individual households. By 1870 there were thirty major lightning rod manufacturers in the United States and perhaps as many as ten thousand traveling lightning rod salesmen, wandering from hamlet to hamlet with their metal contraptions in a wagon, demanding of customers, as in "The Lightning Rod Man," a Herman Melville tale, "Will you order? Will you buy? Think of being a heap of charred offal, like a haltered horse burnt in his stall—and all in one flash!" Sales were made chiefly to rural dwellers, as they had the most intimate acquaintance with the dangers of lightning and knew their wooden barns and outbuildings to be particularly vulnerable. Manufacturers competed with one another by adding unique ornamental elements to their designs, such as weathervanes or colored glass balls.

Thanks to aggressive marketing, the use of lightning rods on private residences increased throughout the nineteenth century, but began to decline by around 1900 as Americans moved from the countryside into towns or cities, where local fire companies were available to douse fires sparked by lightning, and where tall steel-framed buildings, themselves excellent "lightning rods," filled the skyline. Increased mobility in American life also meant that homeowners were less attached to their properties and less willing to invest in lightning rod installation. Fire insurance was often a cheaper option. And although superstitions lingered, people simply became less frightened by the lightning rod salesman's tales of apocalypse after electricity entered the home in more friendly forms, such as the electric light and the telephone. Today, the silent convenience of electricity is so interwoven with our lives, and its benefits so taken for granted, that we tend to feel betrayed when it resorts to its native propensity for "mischief"—when a utility lineman is burned, or a golfer is felled by lightning.

Of course, as electricity became more useful, much was being learned of lightning itself. An important breakthrough was the invention of lightning photography, pioneered by William N. Jennings of Philadelphia, who snapped the world's first photograph of a thunderbolt on September 2, 1882. Another Jennings image of "a streak of real Jersey lightning," taken on August 1, 1885, and later published in *Scientific American*, appeared to have been made looking across the Delaware River from the Philadelphia waterfront, the same perspective from which Franklin observed "thunder-gusts"

more than a century before. The advantages of using photography to study one of the world's fastest natural phenomena were immediately apparent, and researchers who followed Jennings, and improved on his methods, found that panning or rotating the camera during a time exposure showed lightning to be made up of multiple strokes. A seminal invention was the two-lens camera, perfected by British physicist and instrument-maker Charles Vernon Boys. In his design, one lens rotates at high speed around a stationary lens. The latter snaps a normal photograph, while the rotating lens simultaneously captures another image that is smeared over the same film and shows the lightning stroke as it progresses. In this manner, the direction, makeup, and speed of lightning could be analyzed.

"When the first photographs of lightning were made . . . the utter absurdity of the time-honored conception of lightning became apparent," commented *The New York Times* in 1927. "There were no straight lines, no sharp re-entrant angles. Lightning proved to be sinuous and erratic and, what was even more astonishing, branched." The photographs also validated Franklin's contention that lightning moved from earth to sky. "It is what electrical engineers call an 'oscillatory discharge,'" reported the *Times*. "In other words, electricity flashes back and forth until it is spent, like a swinging pendulum that dies down."

Although the difficulties of capturing lightning images in low light, often with wind and rain as impediments, were considerable, the Boys camera allowed researchers to scrutinize closely how discharges of lightning occur. In a typical thundercloud, there are three levels of charge—negative in the middle and positive at the top and bottom. Cloud-to-ground lightning occurs when there is a breakdown between the middle negative and the bottom positive layers that then continues on down to earth. A negative charge builds up at the bottom of a cloud, causing a positive charge to form in the earth below. Because the air between the cloud and the earth is a poor conductor, the opposing charges must build up sufficiently to overload the air's resistance. When that happens, the lightning stroke descends toward the earth in what scientists call a "stepped leader." Stepped leaders are usually about two feet wide and sixty feet long and can grow longer as they approach the ground; as many as forty of these successive branches may be necessary for sky-to-ground contact to be made. As it approaches the earth, the powerful electrical field of the stepped leader causes objects on the ground to emit their own leaders headed upward. Lightning "strikes" when one of these upward leaders contacts the downward leader. The channel that the downward leader has made allows for the more powerful and usually more luminous "return stroke," a wave of ionization that travels up the channel from the earth. Multiple flashes, the oscillating effect, can occur in the same lightning channel, as the second downward stroke, called a "dart leader," is followed by yet another return stroke, weaker than the first. On average, there are four return strokes. All of these flashes happen in less than one ten-thousandth of a second, although because of their intensity they appear to the naked eye to last much longer.

Thunder occurs because lightning's heat (the temperature inside a bolt of lightning can reach 50,000 degrees Fahrenheit, five times that of the sun's surface) violently compresses the surrounding air, creating a shock wave that is heard as a loud crash and then a deep rumble as the sound echoes off other clouds and the earth. Because light travels at 186,000 miles per second, and sound at about 1,100 feet per second, the flash of lightning is visible before the sound of thunder.

Much additional information about lightning has been obtained through our ability, since the dawn of the space age, to gaze down from above at earth's weather. Reports from satellites and manned orbiters tell physicists that at any given time, as many as two thousand thunderstorms are taking place on earth—16 million thunderstorms each year, or 48,000 per day. If each storm is assumed to be about an hour in duration, that means lightning is flashing somewhere on earth about 360,000 times an hour, or one hundred times per second. Researchers have also been able to catalog geographic variances that, in Franklin's time, were known only anecdotally: Cameroon, it is reported, has the world's most thunderstorms, an average of 212 per year, while at the other extreme, the Arctic experiences only about one every ten years. In America, Tampa, on the Florida Gulf Coast, is the lightning mecca, with as many as ninety-four thunderstorms per year, compared with seventy-four in New Orleans, fifty in St. Louis, and about thirty for New York and Philadelphia.

Recent data compiled by the National Oceanic and Atmospheric Administration (NOAA) indicates that in the thirty-six years between 1959 and 1995, there were 3,239 recorded lightning deaths in the United States, 9,818 injuries, and 19,814 property damage reports. As far as meteorological threats, only the devastation from flash-flooding has a higher fatality rate. Still, death by lightning is extremely rare—an average of eighty-seven Americans are killed each year—so much so that being struck by lightning has long been a common metaphor for that which is least likely to occur. It is, however, now as in the era when the lightning rod was invented, a form of sudden death people tend to particularly fear. As Franklin knew, though the chance of being struck by lightning is infinitesimal, the lightning rod's value is that it "may relieve us a hundred times from those painful apprehensions. To make us *safe* is not all its advantage, it is some to make us *easy*."

Of course, no sooner are certain truths about lightning ascertained than it poses yet new mysteries. Still poorly comprehended, for example, is the legendary phenomenon of ball lightning, the Sasquatch of lightning physics. The existence of these ball-shaped bursts of light and energy associated with thunderstorms is supported largely by anecdotal evidence. Some experts have suggested that these may be momentary visual hallucinations caused by close exposure to an actual lightning strike, although there are hundreds of lively accounts describing ball lightning entering homes through windows or doors, coming down (or going up) chimneys, even chasing occupants. A more recently discovered enigma is lightning hidden in the upper atmosphere high above decaying storm systems—upward blasts of brightly colored light fancifully known by the scientists who study them as elves, dwarves, trolls, and sprites. As large as two hundred miles across, they flash only for a micro-second before vanishing. These sightings suggest that thunderstorms exert a powerful influence on the upper atmosphere, but the effect for now remains imperfectly understood.

Franklin's alter ego, Poor Richard, once observed, "A Ship under Sail and a big-bellied woman are the handsomest things that can be seen common." Ships under sail took Franklin many times down the Delaware and far away from Philadelphia. As for his family responsibilities—as husband to Deborah and father to William and Sally—clearly the frequent calls to public duty he heard proved equally if not more enticing. He may have idealized life at home in "dear Philadelphia," as he called it, but after 1757 he spent little time there.

Given that his public life demanded long absences, however, his correspondence shows that he did go to great and often compassionate lengths, using both his influence and his money, to be of help to his children, siblings, nieces, and nephews. With his immediate kin, he compensated by bringing his son, William, then later, William's son, William Temple Franklin, and Sally's young son, Benjamin Franklin Bache, along on his diplomatic travels and postings. But Deborah died in 1774 alone in Philadelphia, unwilling to cross the wide ocean, eventually too sick to even write to her husband in distant London, where his business kept him. His son, William, who as a young man had assisted him with the kite experiment, became, thanks to Franklin's stature in England, the royal governor of New Jersey, but ultimately refused to follow his father into rebellion against the Crown. As a result he was arrested and later deported to England, and the two had virtually no contact for the remainder of Franklin's life. In an age when fathers were also mentors to their sons, and sons were expected to honor them as such, William's decision was to Franklin a hurtful betrayal, and the source of some embarrassment, given Franklin's celebrated status as a patriot.

William was more or less passed over by Franklin in his will. To William's son, William Temple Franklin, he bequeathed a generous sum of money and his papers. Most of his estate he left to his daughter, Sally, and her husband, Richard Bache, with the stipulation that Richard immediately "set free his Negro man Bob." Sally, in taking possession of a valuable portrait of Louis XVI set with 408 diamonds that had been presented to Franklin by the French court, received the fatherly admonition that she never "form any of those diamonds into ornaments . . . and thereby introduce or countenance the expensive, vain, and useless fashion of wearing jewels in this country." Franklin, first and foremost a careful businessman and a meticulous keeper of his own affairs, thus passed on his worldly estate in good order, even establishing a pair of endowments in Boston and Philadelphia for "married artificers under the age of 25 years as have served an apprenticeship in the said town and faithfully fulfilled the duties required in their indentures." In a sense this gesture was a kind of tribute to his own younger self, as he remembered well his early struggle to establish himself as a businessman, husband, and father, while often desperate for funds or trustworthy guidance. "Good apprentices," he, of all people, could honestly conclude, "are most likely to make good citizens."

Franklin, most accounts relate, accepted his last illness and calmly anticipated his own death. If anything did grieve him at the end of his days, it was the knowledge that he would not see the future to which he'd devoted so much thought and contemplation. A dinner guest at Passy in 1779 later recalled that at table Franklin had once "produced a Fly, which had come out of a But of Madeira that Morning, and which by laying in the Sun was restored to Life. The Doctor wish'd, that he cou'd, in like Manner, be bung'd up for fifty Years, and then restored to Life, to behold the flourishing State, in which America wou'd then be." Sharing his vision of the future once with Priestley, Franklin had said, "We may perhaps learn to deprive large Masses of their Gravity & give them absolute Levity, for the sake of easy Transport. Agriculture may diminish its Labour & double its Produce. All Diseases may by sure means be prevented or cured."

Of course nothing would have astonished Franklin more than the rush of electrical innovation in the late nineteenth century. Because of the era's preoccupation with railroads, steamboats, and other steam-driven technology, there was for much of the early nineteenth century scant scientific interest in electricity, and few people saw its potential as a source of power. Limited in Franklin's time to the sparks and charges that could be created with a series of Leydon jars, electricity began to stir in the post-Franklin era of Luigi Galvani and Alessandro Volta, with the discovery of electrical current, a steady flow of electricity. A key development, resulting from nearly simultaneous experiments conducted around 1830 by Michael Faraday in England and by Joseph Henry in the United States, was the discovery that a wire carrying a current of electricity around a piece of iron turns the iron into a magnet, the mechanism upon which the telegraph, telephone, and electric motor were based. Still, an entire century elapsed between Franklin's invention of the lightning rod in 1752 and the second notable application of electrical principles, Samuel Morse's electromagnetic telegraph, which was built on truths already established by Franklin and his generation—that electricity has magnetic powers, and that a charge will travel instantly a considerable distance without losing its strength. After the adoption of Morse's invention and then the laying of the transatlantic cable in 1857, Thomas Edison's electric stock ticker appeared in 1869, followed by his phonograph (1877), light bulb (1879), public generating station (1882), and motion-picture system (1889). In this "era of technological enthusiasm" also came Alexander Graham Bell's telephone, huge urban power grids, and the electrification of trains and streetcars, as well as the pioneering of radio transmissions utilizing electromagnetic radiation in the atmosphere.

Electricity entered and enlarged virtually every aspect of human life, often in the same civic-minded ways Franklin himself would have most approved—from illuminated sidewalks to traffic lights to electric clocks, from innovations that made individual lives safer and more comfortable, to those that sped commerce and enabled human interaction.

The late Enlightenment's view of electricity as a model for addressing social injustice reemerged in the early twentieth century with the Progressive Era's faith that engineers who designed socially beneficial systems might transform other aspects of society. From the time-and-efficiency studies of Frederick W. Taylor to the vision of General Electric's Charles Steinmetz, who saw networked electricity as a means of "industrial cooperation," to such later government projects as the Tennessee Valley Authority and the Rural

Electrification Administration, social betterment was considered a natural extension of engineers' technical efforts, and of electricity itself.

Surely one development that would please Franklin the electrician but perhaps worry Franklin the pragmatist is the extent to which mankind today has become not simply reliant but utterly dependent on electricity—the "subtile fluid" he often feared had no discoverable purpose.

It should perhaps be no surprise that toward the end of a life spent investigating earthly enigmas, Franklin's attention would turn finally to that greatest of unknowables, the afterlife. In his letters he had often used that locale as a pretext for humorous pronouncements (John Adams recalled that Franklin once joked, "We are all invited to a great entertainment. Your carriage comes first to the door; but we shall all meet there") or as a paradise in which his love of certain inaccessible French ladies would at last be requited. One of the longtime objects of his affection, Madame Brillon de Jouy, joined Franklin in imagining a place in the hereafter where "Mr. Mesmer will be contented with playing on the armonica with-

out boring us with electric fluid." To his friend George Whatley he wrote, "I look upon death to be as necessary to our constitution as sleep. We shall rise refreshed in the morning." And to the clergyman Ezra Stiles, who queried him about his religious views, Franklin affirmed his faith in God but admitted to doubts about the divinity of Jesus, "tho' it is a question I do not dogmatize upon, having never studied it, and think it needless to busy myself with now, when I expect soon an opportunity of knowing the truth with less trouble."

Franklin, as he grew older, evinced a more sincere curiosity about the fate of souls. His practical, order-seeking mind, perceiving that so much wickedness in life went unpunished, resolved that some correcting force, a kind of cosmic balance sheet, must exist, so that in "a future State . . . all that here appears wrong shall be set right, [and] all that is crooked made straight." And, having observed over the course of his long life that among nature's many admirable perfections was that it wasted nothing, he developed ultimately a more or less ecological view of the soul's transformation.

"I cannot suspect the Annihilation of Souls," he wrote, "or believe, that [God] will suffer the daily Waste of Millions of Minds ready made that now exist, and put himself to the continual Trouble of making new ones. I believe," he concluded with prophetic accuracy, "I shall, in some Shape or other, always exist."

ACKNOWLEDGMENTS

ONE OF THE PLEASURES OF WRITING A BOOK ABOUT BENJAMIN Franklin is that his generosity of spirit endures among the scholars and archivists who best know his life and work. I wish to thank Ellen Cohn, Kate Ohno, and Jennifer Macellaro of the Papers of Benjamin Franklin at Yale University for their assistance and advice, and also for making me feel welcome in New Haven. Claude-Anne Lopez, a former editor of the Papers, introduced me to the collection and the staff there, and has been a good friend throughout this project. Roy Goodman at the American Philosophical Library in Philadelphia has likewise been a constant source of support. David Rhees at the Bakken Library and Museum in Minneapolis was instrumental in my attending an international conference on the lightning rod at the Bakken in fall 2002, at which I was introduced to many sides of the lightning rod story as well as several individuals who have graciously lent me their help, especially

Professor Roderick Home of the University of Melbourne, Philip Krider, from the Institute of Atmospheric Physics at the University of Arizona, and Graydon Aulich of the Langmuir Laboratory for Atmospheric Research at the New Mexico Institute of Mining and Technology.

Mary Beth Norton, professor of history at Cornell University, and John Beihan of Loyola College in Baltimore gave valuable suggestions about the manuscript. Paola Bertucci of the University of Bologna was kind enough to send me a copy of *Electric Bodies: Episodes in the History of Medical Electricity*, an anthology she edited with Giuliano Pancaldi. Jan Gross, James Sanders, Robert Sietsema, and Lianne Smith offered early encouragement for the book; Mary Gibney and Stephanie Steiker provided valuable assistance. Historian David Levering Lewis has been generous with his support and friendship.

In addition to the Papers of Benjamin Franklin at Yale University and the American Philosophical Library in Philadelphia, I was able to perform valuable research at the Harvard University Library Archives, the Massachusetts Historical Society, the New York Public Library Rare Book Division, and the Bobst Library at New York University.

I. Bernard Cohen, for many years a professor of the history of science at Harvard University, was one of the first American historians to fully appreciate Franklin's scientific side and to emphasize the importance of the lightning rod. His *Franklin and Newton, Science and the Founding Fathers* and *Benjamin Franklin's Science*, among other writings, are indispensable to anyone who wishes to comprehend Franklin and eighteenth-century natural philosophy. Three books by historian Joseph J. Ellis—*After the Revolution, Founding Brothers*, and *Passionate Sage: The Character and Legacy of John Adams*—guided my understanding of the leading figures of the founding generation. *Worlds of Wonder, Days of Judgment* by David Hall was instrumental for its views of belief and custom in colonial New England, and Andrew Dickson White's *History of the Warfare of Science with Theology in Christendom* was a fruitful source on the church's reaction to Franklin's invention. Also helpful were Jessica Riskin's superb *Science in the Age of Sensibility;* Claude-Anne Lopez's

Mon Cher Papa and *The Private Franklin* (the latter title written with Eugenia Herbert); and the seminal *Rebels and Gentlemen* by Carl and Jessica Bridenbaugh, which remains perhaps the best introduction to the world of Franklin's Philadelphia.

Production editor Janet Wygal, copy editor Jonathan Thomas, cover designer Gabrielle Bordwin, book designer Pei Loi Koay, and editorial assistant Julia Cheiffetz were essential to the book's publication.

Finally, I wish to thank Elizabeth Sheinkman and Geri Thoma at the Elaine Markson Literary Agency for their hard work and support; Scott Moyers, who believed in the book when it was only a rumble in the distance; and my editor at Random House, David Ebershoff, whose dedicated efforts helped bring the manuscript to completion.

BIBLIOGRAPHY

BOOKS AND MAJOR SOURCES

Aldridge, Alfred Owens. *Benjamin Franklin and Nature's God*. Durham, N.C.: Duke University Press, 1967.

————. *Franklin and His French Contemporaries*. New York: New York University Press, 1957.

Alsop, Susan Mary. *Yankees at the Court: The First Americans in Paris*. New York: Washington Square Press, 1985 (originally published 1982).

Arago, François. *Meteorological Essays*. London: Longman, Brown, Green & Longmans, 1855.

Baker, Keith Michael. *Condorcet: From Natural Philosophy to Social Mathematics*. Chicago: University of Chicago Press, 1975.

Baltzell, E. Digby. *Puritan Boston and Quaker Philadelphia*. New York: The Free Press, 1979.

Becker, Carl L. *Benjamin Franklin: A Biographical Sketch*. Ithaca, N.Y.: Cornell University Press, 1946.

————. *The Heavenly City of the Eighteenth-Century Philosophers*. New Haven: Yale University Press, 1932.

Bell, Whitfield J., Jr. *Early American Science: Needs and Opportunities*. Williamsburg, Va.: Institute of Early American History and Culture, 1955.

Benjamin, Park. *The Age of Electricity*. New York: Scribners, 1886.

Berkeley, Edward, and Dorothy Smith Berkeley. *Dr. John Mitchell: The Man Who Made the Map*. Chapel Hill: University of North Carolina Press, 1974.

Berkin, Carol. *A Brilliant Solution: Inventing the American Constitution*. New York: Harcourt Brace, 2002.

Bertucci, Paola, and Giuliano Pancaldi, editors. *Electric Bodies: Episodes in the History of Medical Electricity*. Bologna, Italy: Bologna Studies in the History of Science, Department of Philosophy, University of Bologna, 2001.

Bishop, J. Leander. *A History of American Manufacturers from 1680–1860*. Philadelphia: Edward Young & Co., 1868.

Bodzin, Eugene Saul. *The American Popular Image of Benjamin Franklin 1790–1868*. Ann Arbor, Mich.: University Microforms, 1970.

Boorstin, Daniel. *The Lost World of Thomas Jefferson*. Boston: Beacon Press, 1948.

Bowen, Catherine Drinker. *The Most Dangerous Man in America: Scenes from the Life of Benjamin Franklin*. Boston: Little, Brown and Company, 1974.

Brands, H. W. *The First American: The Life and Times of Benjamin Franklin*. New York: Doubleday, 2000.

Braudy, Leo. *The Frenzy of Renown: Fame and Its History*. New York: Oxford University Press, 1986.

Breitwieser, Mitchell R. *Cotton Mather and Benjamin Franklin: The Price of Representative Personality*. New York: Columbia University Press, 1984.

Brett-James, Norman G. *The Life of Peter Collinson*. London: Edgar G. Dunstan & Co., 1928.

Bridenbaugh, Carl and Jessica. *Rebels and Gentlemen: Philadelphia in the Age of Franklin*. New York: Oxford University Press, 1965.

Buffon, George Louis Leclerc. *Correspondence de Buffon de 1729 à 1788*. Paris: Volume I, 1885.

———. *Des Époques de la Nature*. Paris: Éditions Rationalistes, 1971.

Campbell, James. *Recovering Benjamin Franklin: An Exploration of a Life of Science and Service*. Chicago: Open Court, 1999.

Carr, William G. *The Oldest Delegate: Franklin in the Constitutional Convention*. Newark: University of Delaware Press, 1990.

Castle, Terry. *The Female Thermometer: Eighteenth Century Culture and the Invention of the Uncanny*. New York: Oxford University Press, 1995.

Clark, Ronald. *Benjamin Franklin*. New York: Random House, 1983.

Cohen, Claudine. *The Fate of the Mammoth: Fossils, Myth, and History*. Chicago: University of Chicago Press, 2002.

Cohen, I. Bernard. *Benjamin Franklin: Scientist and Statesman*. New York: Charles Scribners, 1972.

———. *Benjamin Franklin's Experiments*. Cambridge, Mass.: Harvard University Press, 1941.

———. *Benjamin Franklin's Science*. Cambridge, Mass.: Harvard University Press, 1990.

———. *Franklin and Newton*. Cambridge, Mass.: Harvard University Press, 1966.

———. *Science and the Founding Fathers: Science in the Political Thought of Thomas Jefferson, Benjamin Franklin, John Adams and James Madison*. New York: W. W. Norton and Company, 1995.

————. *Some Early Tools of American Science.* Cambridge, Mass.: Harvard University Press, 1950.

Colden, Cadwallader. *The Letters and Papers of Cadwallader Colden.* New York: New York Historical Society, 1920.

Collinson, Peter. *Forget Not Mee and My Garden.* Correspondence of Peter Collinson, edited and introduced by Alan Armstrong. Philadelphia: American Philosophical Society, 2002.

Crabtree, Adam. *From Mesmer to Freud: Magnetic Sleep and the Roots of Psychological Healing.* New Haven: Yale University Press, 1993.

Cutler, William P. and Julia. *Life, Journals, and Correspondence of Rev. Manasseh Cutler.* Athens: Ohio University Press, 1987.

Darnton, Robert. *George Washington's False Teeth: An Unconventional Guide to the Eighteenth Century.* New York: W. W. Norton & Company, 2003.

————. *Mesmerism and the End of the Enlightenment in France.* Cambridge, Mass.: Harvard University Press, 1968.

Day, W. H. *Lightning Rods.* Ottawa, Ontario: Canada Department of Agriculture, Bulletin 220, March 1914.

Deacon, Margaret. *Scientists and the Sea 1650–1900: A Study of Marine Science.* New York: Academic Press, 1971.

DePauw, Linda, editor. *Documentary History of the First Federal Congress 1789–1791,* Volume 12. Baltimore: Johns Hopkins University, 1994.

Dibner, Bern. *Early Electrical Machines.* Norwalk, Conn.: The Burndy Library, 1957.

————. *Galvani-Volta: A Controversy That Led to the Discovery of Useful Electricity.* Norwalk, Conn.. The Burndy Library, 1952.

Duniway, Clyde A. *Development of Freedom of the Press in Massachusetts.* New York: Burt Franklin, 1969.

Edwards, Jonathan. *Works.* New York: G. & C. & H. Carvill, 1830.

Ellis, Joseph J. *After the Revolution: Profiles of Early American Culture.* New York: W. W. Norton and Company, 1979.

————. *Founding Brothers: The Revolutionary Generation.* New York: Alfred A. Knopf, 2001.

————. *Passionate Sage: The Character and Legacy of John Adams.* New York: W. W. Norton and Company, 1993.

Emery, Edwin. *The Press and America: An Interpretive History of the Mass Media.* Englewood Cliffs, N.J.: Prentice-Hall, 1954.

Fara, Patricia. *An Entertainment for Angels: Electricity in the Enlightenment.* Cambridge, U.K.: Icon Books Ltd., 2002.

Farand, Max, editor. *The Records of the Federal Convention of 1787.* New Haven: Yale University Press, 1911.

Fay, Bernard. *Franklin: The Apostle of Modern Man.* Boston: Little, Brown, 1929.

Fellows, Otis E., and Stephen F. Milliken. *Buffon.* New York: Twayne Publishers, 1972.

Fleming, Thomas. *The Man Who Dared the Lightning: A New Look at Benjamin Franklin.* New York: William Morrow and Company, 1971.

Fonvielle, W. De. *Thunder and Lightning.* New York: Charles Scribners, 1875.

Forbes, R. B. *Shipwreck by Lightning.* Boston: Sleeper & Rogers, 1853.

Fox, Hingston. *Dr. John Fothergill and His Friends: Chapters in Eighteenth Century Life.* London: Macmillan & Co. Ltd., 1919.

Frangsmyr, Tone, editor. *Linnaeus: The Man and His Work.* Berkeley: University of California Press, 1983.

Franklin, Benjamin. *The Autobiography of Benjamin Franklin.* Leonard W. Larabee, editor. New Haven: Yale University Press, 1964.

———. *Experiments and Observations on Electricity, Made at Philadelphia in America.* London: E. Cave, 1751.

———. *The Papers of Benjamin Franklin,* Volumes 1–37. Leonard Labaree et al., editors. New Haven: Yale University Press, 1959.

———. *The Writings of Benjamin Franklin.* Albert Henry Smyth, editor. New York: Macmillan, 1907; reprinted, New York: Haskell House, 1970.

Franklin, Phyllis. *Show Thyself a Man: A Comparison of Benjamin Franklin and Cotton Mather.* The Hague: Mouton, 1969.

Franklin, William Temple. *Memoirs of the Life and Writings of Benjamin Franklin.* Three volumes. London: Henry Colburn, 1818.

Furbank, P. N. *Diderot.* London: Secker & Warburg, 1992.

Gaskell, T. F. *The Gulf Stream.* New York: John Day Co., 1973.

Gay, Peter. *The Enlightenment: An Interpretation: The Science of Freedom.* New York: W. W. Norton and Company, paperback reissue, 1996.

Gillespie, Charles. *Science and Polity in France at the End of the Old Regime.* Princeton, N.J.: Princeton University Press, 1980.

Golde, R. H. *Lightning Protection.* London: Edward Arnold Publishers, 1973.

Goodman, Nathan G., editor. *The Ingenious Dr. Franklin: Selected Scientific Letters of Benjamin Franklin.* Philadelphia: University of Pennsylvania Press, 1931.

Gorst, Martin. *Measuring Eternity: The Search for the Beginning of Time.* New York: Broadway Books, 2001.

Hall, David D. *Worlds of Wonder, Days of Judgment: Popular Religious Belief in Early New England.* Cambridge, Mass.: Harvard University Press, 1990.

Hamblyn, Richard. *The Invention of Clouds: How an Amateur Meteorologist Forged the Language of the Skies.* New York: Farrar, Straus and Giroux, 2001.

HaNian-Sheng, Huang. *Benjamin Franklin in American Thought and Culture 1790–1990.* Philadelphia: American Philosophical Society, 1994.

Harris, W. Snow. *Observations of the Effects of Lightning on Floating Bodies.* London: W. Nicol, 1823.

———. *On the Nature of Thunderstorms, and on the Means of Protecting Buildings and Shipping against the Destructive Effects of Lightning.* London: J. W. Parker, 1843.

Heilbron, J. L. *Electricity in the 17th and 18th Centuries.* Berkeley: University of California Press, 1979.

Henry, Alfred J. *Loss of Life in the United States by Lightning.* Washington, D.C.: Government Printing Office, 1901.

Hindle, Brooke. *The Pursuit of Science in Revolutionary America, 1735–1789.* Chapel Hill: University of North Carolina Press, 1956.

Hindle, Brooke, editor. *The History of Science Selections from ISIS.* New York: Science History Publications, 1976.

Isaacson, Walter. *Benjamin Franklin: An American Life.* New York: Simon & Schuster, 2003.

Jefferson, Thomas. *Notes on Virginia.* Chapel Hill: University of North Carolina Press, 1994.

Kendrick, T. D. *The Lisbon Earthquake.* London: Methuen and Co. Ltd., 1956.

Kraus, Michael. *The Atlantic Civilization: Eighteenth Century Origins.* Ithaca, N.Y.: Cornell University Press, 1966 (originally published by the American Historical Association, 1949).

Lambert, Frank. *Inventing the "Great Awakening."* Princeton, N.J.: Princeton University Press, 1999.

Leonard, Thomas C. *The Power of the Press: The Birth of American Political Reporting.* New York: Oxford University Press, 1986.

Lepore, Jill. *The Name of War: King Philip's War and the Origins of American Identity.* New York: Alfred A. Knopf, 1998.

Lokken, Roy N., editor. *Meet Dr. Franklin.* Philadelphia: Franklin Institute Press, 1988.

Lopez, Claude-Anne. *Benjamin Franklin's "Good House": The Story of Franklin Court.* Washington, D.C.: Division of Publications, National Park Service, 1981.

———. *Mon Cher Papa: Franklin and the Ladies of Paris.* New Haven: Yale University Press, 1966.

———. *My Life with Franklin.* New Haven: Yale University Press, 2002.

Lopez, Claude-Anne, and Eugenia W. Herbert. *The Private Franklin: The Man and His Family.* New York: W. W. Norton and Company, 1975.

Love, W. De Loss, Jr. *The Fast and Thanksgiving Days of New England.* Boston: Houghton Mifflin, 1895.

Mather, Cotton. *Magnalia Christi Americana, Books I and II,* edited by Kenneth B. Murdock. Cambridge, Mass.: Harvard University Press, 1977.

———. *The Angel of Bethesda,* edited by Gordon Jones. Barre, Mass.: Barre Publishing, 1972.

Maurois, André. *The Miracle of France.* New York: Farrar, Straus and Cudahy, 1948.

McAdie, Alexander. *Protection from Lightning.* Washington, D.C.: U.S. Weather Bureau, 1894.

McCullough, David. *John Adams.* New York: Simon & Schuster, 2001.

McEachron, K. B., with Kenneth G. Patrick. *Playing with Lightning.* New York: Random House, 1940.

McGowan, Christopher. *The Dragon Seekers.* Cambridge, Mass.: Perseus Publishing, 2001.

Middlekauf, Robert. *Benjamin Franklin and His Enemies.* Berkeley: University of California Press, 1996.

Miller, D. C. *Sparks, Lightning, Cosmic Rays: An Anecdotal History of Electricity.* New York: Macmillan, 1939.

Miller, Perry. *The New England Mind: From Colony to Province.* Cambridge, Mass.: Harvard University Press, 1953.

Millikan, Robert. *Electrons, Protons, Photons, Neutrons, and Cosmic Rays.* Chicago: University of Chicago Press, 1935.

Morgan, Edmund. *Benjamin Franklin.* New Haven: Yale University Press, 2002.

————. *The Gentle Puritan: A Life of Ezra Stiles, 1727–1795*. New Haven: Yale University Press, 1962.

Neiman, Susan. *Evil in Modern Thought: An Alternative History of Philosophy*. Princeton, N.J.: Princeton University Press, 2002.

Newton, Isaac. *Optiks*. London: Smith & Walford, 1704.

Nicholson, William. *An Introduction to Natural Philosophy, Volume 2, 3rd edition*. London: R. Johnson, Publisher, 1790.

Niklaus, Robert. *A Literary History of France, Volume III: The Eighteenth Century 1715–1789*. London: Benn, 1970.

Nollet, Jean-Antoine. *Programme ou Idée Générale d'un Cours de Physique Expérimentale*. Paris: 1738.

Norton, Mary Beth. *In the Devil's Snare: The Salem Witchcraft Crisis of 1692*. New York: Alfred A. Knopf, 2002.

Parton, James. *Life and Times of Benjamin Franklin, Volume 2*. New York: Mason Brothers, 1864.

Peale, Rembrandt. *Account of the Skeleton of the Mammoth*. London: E. Lawrence, 1802.

Pepper, William. *The Medical Side of Benjamin Franklin*. New York: Argosy-Antiquarian, 1970.

Petrey, Sandy, editor. *The French Revolution 1789–1989: Two Hundred Years of Rethinking*. Lubbock: Texas Tech University Press, 1989.

Pick, Daniel. *Svengali's Web: The Alien Enchanter in Modern Culture*. New Haven: Yale University Press, 2000.

Pliny the Elder. *Natural History (Selections)*. John F. Healy, translator. London: Penguin, 1991.

Prager, Frank D., editor. *Autobiography of John Fitch*. Philadelphia: American Philosophical Society, 1976.

Priestley, Joseph. *The History and Present State of Electricity*. London: C. Bathurst, 4th edition, 1775.

Prince, Thomas. *Earthquakes the Works of God and Tokens of His Just Displeasure . . . on Occasion of the late Dreadful Earthquake which Happened on the 18th of Nov. 1755*. Boston: D. Fowle & Z. Fowle, 1755.

Ramsey, David. *The History of the American Revolution*. Edited by Lester H. Cohen. Indianapolis: Liberty Classics, 1990.

Raynal, Abbé. *A Philosophical and Political History of the Settlements and Trade of the Europeans in the East and West Indies*. Translated from the French. Edinburgh: Caddel & Balfour, 1776.

Richards, William C. *Electron, or Pranks of the Moden Puck*. New York: D. Appleton & Co., 1858.

Riskin, Jessica. *Science in the Age of Sensibility: The Sentimental Empiricists of the French Enlightenment*. Chicago: University of Chicago Press, 2002.

Robespierre, Maximilien. *Discours et rapports à la convention, par Robespierre*. Paris: Union Générale d'Éditions, 1965.

Rossiter, Clinton. *Seedtime of the Republic: The Origin of the American Tradition of Political Liberty*. New York: Harcourt Brace, 1953.

Rush, Benjamin. *Autobiography of Benjamin Rush*. George W. Corner, editor. Westport, Conn.: Greenwood Press, 1970.

Sagan, Carl. *The Demon-Haunted World: Science as a Candle in the Dark.* New York: Random House, 1995.

Schama, Simon. *Citizens.* New York: Alfred A. Knopf, 1989.

Schecter, Barnet. *The Battle for New York, the City at the Heart of the American Revolution.* New York: Penguin Books, 2002.

Schneller, Rachel E. "The Extra, Groundless Machinations of Men: Cotton Mather, William Douglass, and the Smallpox Inoculation Controversy in Boston, 1721–1722." Harvard University A.B. thesis, 1998.

Schofield, Robert E. *Mechanism and Materialism: British Natural Philosophy in an Age of Reason.* Princeton, N.J.: Princeton University Press, 1970.

Schonland, Basil F. *The Flight of Thunderbolts.* Oxford, U.K.: Clarendon Press, 1964.

Seeger, Raymond John. *Benjamin Franklin: New World Physicist.* Oxford, U.K.: Pergamon Press, 1973.

Shute, Michael N., editor. *The Scientific Work of John Winthrop.* New York: Arno Press, 1980.

Silverman, Kenneth, editor. *Selected Letters of Cotton Mather.* Baton Rouge: Louisiana State University Press, 1971.

Singer, Stanley. *The Nature of Ball Lightning.* New York: Plenum Press, 1971.

Solberg, Winton V., editor. *Cotton Mather: The Christian Philosopher.* Urbana: University of Illinois Press, 1994.

Sonneck, O.G.T. *Suum Cuiques: Essays on Music.* New York: G. Schirmer, 1916.

Starkey, Marion L. *The Devil in Massachusetts: A Modern Enquiry into the Salem Witch Trials.* New York: Alfred A. Knopf, 1949.

Stearns, Raymond P. *Science in the British Colonies of America.* Carbondale: University of Illinois Press, 1970.

Stifler, James Madison. *The Religion of Benjamin Franklin.* New York: D. Appleton and Company, 1925.

Still, Alfred. *The Soul of Amber: The Background of Electrical Science.* New York: Murray Hill Books, 1944.

Stout, Harry S. *The Divine Dramatist: George Whitefield and the Rise of Modern Evangelism.* Grand Rapids, Mich.: William B. Eerdmans Publishing, 1991.

Struik, Dirk J. *Yankee Science in the Making.* New York: Collier Books, 1962.

Swem, E. G., editor. *Brothers of the Spade: Correspondence of Peter Collinson and John Custus, 1734–1746.* Barre, Mass.: Barre Gazette, 1957.

Taylor, Alan. *American Colonies: The Settling of North America.* New York: Viking Penguin, 2001.

Thomas, Keith. *Religion and the Decline of Magic.* New York: Scribners, 1971.

Tolles, Frederick B. *Meeting House and Counting House: The Quaker Merchants of Colonial Philadelphia 1682–1763.* New York: W. W. Norton and Company, 1963. Originally published in 1948 by University of North Carolina Press.

Tomlinson, W. *The Thunder-Storm.* London: Society for Promoting Christian Knowledge, 1848.

Tourtellot, Arthur B. *Benjamin Franklin: The Shaping of Genius: The Boston Years.* Garden City, N.Y.: Doubleday, 1977.

Trowbridge, John. *The Advance in Electricity Since the Time of Franklin.* Cambridge, Mass.: Harvard University Press, 1922.

Tucker, Tom. *Bolt of Fate: Benjamin Franklin and His Electric Kite Hoax.* New York: Public Affairs, 2003.

Uman, Martin A. *All About Lightning.* New York: Dover Publications, 1986.

Van Doren, Carl. *Benjamin Franklin* (Penguin reprint). Originally published, New York: Viking Press, 1938.

———. *Jane Mecom: Franklin's Favorite Sister.* New York: Viking Press, 1950.

Vellay, Charles. *Robespierre et le Procès du Paratonerre.* Le Puy, France: Imprimerie Peyriller, 1909.

Viemeister, Peter E. *The Lightning Book.* Garden City, N.Y.: Doubleday, 1961.

Wainwright, Nicholas B. *George Croghan: Wilderness Diplomat.* Chapel Hill: University of North Carolina Press, 1959.

Walters, Kerry S. *Benjamin Franklin and His Gods.* Urbana: University of Illinois Press, 1999.

Warden, G. B. *Boston: 1689–1776.* Boston: Little, Brown, 1970.

Westcott, Thompson. *A History of Philadelphia.* Philadelphia: Pawson & Nicolson, 1886.

White, Andrew Dickson. *History of the Warfare of Science with Theology in Christendom.* New York: Dover Publications, 1960.

Whyte, Adam. *The All-Electric Age.* London: Constable & Co., Ltd., 1922.

Wightman, Joseph M. *A Companion to Electricity.* Boston: Samuel N. Dickinson, 1843.

Wolf, A. *A History of Science, Technology, and Philosophy in the Eighteenth Century.* London: George Allen, 1938.

Wood, Gordon S. *The American Revolution: A History.* New York: Modern Library, 2002.

Wright, Esmond. *Franklin of Philadelphia.* Cambridge, Mass.: Harvard University Press, 1986.

Wright, Esmond, editor. *Benjamin Franklin: A Profile.* New York: Hill and Wang, 1970.

Zall, P. N. *Ben Franklin Laughing: Anecdotes from Original Sources.* Berkeley: University of California Press, 1980.

Zim, Herbert S. *Lightning and Thunder.* New York: William Morrow & Co., 1952.

ARTICLES

Aldridge, A. O. "Franklin and the Ghostly Drummer of Tedworth." *William and Mary Quarterly Historical Magazine,* Series 3, Volume 7, 1950.

Aldridge, Alfred Owen. "Benjamin Franklin and Jonathan Edwards on Lightning and Earthquakes." In Hindle, Brooke, editor, *The History of Science Selections from ISIS.* New York: Science History Publications, 1976.

Anonymous. "On Pictures Taken by Lightning." *Appleton's Journal of Popular Literature, Science, and Art,* October 23, 1869.

Bender, Bert A. "Let There Be (Electric) Light! The Image of Electricity in American Writing." *Arizona Quarterly,* Volume 34, no. 1, 1978.

Bercovitch, Sacvan. "The Puritan Vision of the New World." In *The Columbia Literary History of the United States,* Emory Elliott, editor. New York: Columbia University Press, 1988.

Biggs, E. Power. "The Story of Benjamin Franklin and the Armonica." Notes to a Concert of Chamber Music by Benjamin Franklin and Wolfgang Amadeus Mozart. Cambridge, Mass.: Massachusetts Institute of Technology, April 11, 1956.

Brasch, Frederick E. "John Winthrop, America's First Astronomer, and the Science of His Period." *Astronomical Society of the Pacific*, Volume 28, no. 165, 1916.

Clark, Charles E. "Boston and the Nurturing of Newspapers: Dimensions of the Cradle, 1690–1741." *New England Quarterly*, Volume 64, no. 2, 1991.

Emery, Allan Moore. "Melville on Science: 'The Lightning Rod Man.' " *New England Quarterly*, Volume 56, no. 4, 1983.

Franklin, Benjamin. "A Letter from Dr. Benjamin Franklin to Mr. Alphonsus le Roy . . . containing sundry Maritime Observations." *Transactions of the American Philosophical Society*, Volume 2, no. 38, 1786.

Gold, Joel J. "Dinner at Doctor Franklin's." *Modern Philology*, Volume 75, no. 4, 1978.

Griffith, John. "Franklin's Sanity and the Man Behind the Masks." In Lemay, J. A. Leo, editor, *The Oldest Revolutionary: Essays on Benjamin Franklin*. Philadelphia: University of Pennsylvania Press, 1976.

Hales, Stephen. "Some Consideration on the Causes of Earthquakes." *Philosophical Transactions*, Volume 46, no. 4, 1749–1750.

Haraszti, Zoltan. "Young John Adams on Franklin's Iron Points." *ISIS*, Volume 41, no. 2, 1950.

Heathcote, N. H. de V. "Franklin's Introduction to Electricity." *ISIS*, Volume 46, Part 1, no. 143, 1955.

Hindle, Brooke. "The Quaker Background and Science in Colonial Philadelphia." In Hindle, Brooke, editor, *The History of Science Selections from ISIS*. New York: Science History Publications, 1976.

Home, R. W. "Nollet and Boerhaave: A Note on Eighteenth-Century Ideas about Electricity and Fire." *Annals of Science*, Volume 36, no. 2, 1979.

Hornberger, Theodore. "The Science of Thomas Prince." *New England Quarterly*, Volume 9, no. 1, 1936.

Houston, Jourdan. "When the Great Earthquake Struck New England." *American Heritage*, August/September, 1980.

Huet, Marie-Helene. "Thunder and Revolution: Franklin, Robespierre, Sade." In Petrey, Sandy, editor, *The French Revolution 1789–1989: Two Hundred Years of Rethinking*. Lubbock: Texas Tech University Press, 1989.

James, A.J.L. "Davy in the Dockyard: Humphry Davy, the Royal Society, and the Electro-Chemical Protection of Copper Sheathing on His Majesty's Ships in the 1820s." *Physis*, Volume 29, no. 1, 1992.

James, A.J.L., and J. V. Field. "Frankenstein and the Spark of Being." *History Today*, Volume 44; issue 9, September 1994.

Jennings, W. N. "Lightning Photography." *Scientific American*, September 5, 1885.

Kerr, Joann P. "What Good Is a New-Born Baby?" *American Heritage*, December 1973.

Lemay, J.A.L. "Franklin and Kinnersley." In Hindle, Brooke, editor, *The History of Science Selections from ISIS*. New York: Science History Publications, 1976.

Lemay, Leo. "Benjamin Franklin: Universal Genius, the Renaissance Man in the Eighteenth Century." Paper read at a Clark Library Seminar, October 9, 1976, Los Angeles, University of California, William A. Clark Memorial Library.

Lipinski, Edward R. "Lightning Rods Require Expert Hand to Install." *New York Times*, March 5, 1995.

McEachron, K. B. "Lightning Protection Since Franklin's Day." *Journal of the Franklin Institute*, Volume 253, no. 5, 1952.

———. "Lightning to the Empire State Building." *Journal of the Franklin Institute*, Volume 227, no. 2, 1939.

Millikan, Robert A. "Benjamin Franklin as a Scientist." In Lokken, Roy, editor, *Meet Dr. Franklin*. Philadelphia: Franklin Institute Press, 1988.

———. "Franklin's Discovery of the Electron." *American Journal of Physics*, Volume 16, no. 5, 1948.

Mulford, Carla. "Figuring Benjamin Franklin in American Cultural Memory." *New England Quarterly*, Volume 72, no. 3, 1999.

Mukerji, Chandra. "Reading and Writing with Nature: A Materialist Approach to French Formal Gardens." In Brewer, John, and Roy Porter, editors, *Consumption and the World of Goods*. London: Routledge Books, 1993.

Nickels, Cameron C. "Franklin's Poor Richard's Almanacs: 'The Humblest of His Labors.'" In Lemay, J. A. Leo, editor, *The Oldest Revolutionary: Essays on Benjamin Franklin*. Philadelphia: University of Pennsylvania Press, 1976.

Nicolella, Michael. "Glass Armonica: A Brief History." *Arts4All Newsletter*, Volume I, Issue 4, August 16, 1999.

Norinder, Harald. "Experimental Lightning Research." *Journal of the Franklin Institute*, Volume 253, no. 5, 1952.

Ohline, Howard A. "Slavery, Economics, and Congressional Politics, 1790." *The Journal of Southern History*, Volume 46, no. 3, 1980.

Rothstein, Edward. "Defining Evil in the Wake of 9/11." *New York Times*, October 5, 2002.

———. "Playing on Glass." *New York Times*, January 15, 1984.

Schaffer, Simon. "The Consuming Flame: Electrical Showmen and Tory Mystics in the World of Goods." In Brewer, John, and Roy Porter, editors, *Consumption and the World of Goods*. London: Routledge Books, 1993.

———. "Natural Philosophy and Public Spectacle in the Eighteenth Century." *History of Science*, Volume 21, Part 1, no. 51, 1983.

Schlesinger, Arthur M. "An American Historian Looks at Science and Technology." In Hindle, Brooke, editor, *The History of Science Selections from ISIS*. New York: Science History Publications, 1976.

Schonland, Basil F. "The Work of Benjamin Franklin on Thunderstorms and the Development of the Lightning Rod." *Journal of the Franklin Institute*, Volume 253, no. 5, 1952.

Scott, Gale. "With Towers Gone, Area May Be Vulnerable to Lightning." *New York Times*, September 3, 2002.

Silverman, Kenneth. "From Cotton Mather to Benjamin Franklin." In Emory, Elliott, editor, *The Columbia Literary History of the United States*. New York: Columbia University Press, 1988.

Simpson, George Gaylord. "The Beginnings of Vertebrate Paleontology in North America." *Proceedings of the American Philosophical Society*, Volume 86, no. 1, 1942.

Singer, Dorothea Waley. "Sir John Pringle and His Circle." *Annals of Science*, Volume 6, nos. 2 and 3, 1948–1950.

Stearns, Raymond. "Remarks upon the Introduction of Inoculation for Smallpox in England." *Bulletin of the History of Medicine*, Volume 24, 1950.

Stephens, Brad. "A List of Benjamin Franklin's Outstanding Achievements." *The Benjamin Franklin Gazette*, February 1943.

Stukeley, William. "[Letter] to the President, the Philosophy of Earthquakes." *Philosophical Transactions*, Volume 46, no. 497, 1749–1750.

Tilton, Eleanor M. "Lightning Rods and the Earthquake of 1755." *New England Quarterly*, Volume 13, no. 1, 1940.

Tolles, Frederick B. "Philadelphia's First Scientist: James Logan." In Hindle, Brooke, editor, *The History of Science Selections from ISIS*. New York: Science History Publications, 1976.

Trowbridge, John. "Lightning and Lightning Rods." *Atlantic Monthly*, Volume 36, July 1875.

Updike, John. "Many Bens." *The New Yorker*, February 22, 1988.

Van Marum, Martin. "On the Theory of Franklin, According to Which Electrical Phenomena Are Explained by a Single Fluid." *Annals of Philosophy*, Volume 16, July–December 1820.

Vellay, Charles. "Robespierre et le Procès du Paratonnerre (1780–1784)." *Annales Révolutionnaires*, January–March 1909, and April–June 1909. Published separately, Le Puy: Imprimerie Peyriller, Rouchon & Gamon, 1909.

Wetering, John F. Van de. "God, Science, and the Puritan Dilemma." *New England Quarterly*, Volume 38, no. 4, 1965.

Wetering, Maxine Van de. "A Reconsideration of the Inoculation Controversy." *New England Quarterly*, Volume 58, no. 1, 1985.

Wilford, John Noble. "Lightning Rods: Franklin Had It Wrong." *New York Times*, July 14, 1983.

———. "NASA Readies New Assault on Still-Mysterious Lightning." *New York Times*, July 14, 1987.

Winthrop, John. "An Account of the Earthquake Felt in New England." *Philosophical Transactions*, Volume 50, 1757–58.

———. Appendix to a Lecture on Earthquakes: "Concerning the Operation of Electrical Substance in Earthquakes; and the Effects of Iron Points." Cambridge, Mass.; December 1755.

Wright, Esmond. "Benjamin Franklin: 'The Old England Man.'" In Klein, Randolph S., editor, *Science and Society in Early America: Essays in Honor of Whitfield J. Bell, Jr.* Philadelphia: American Philosophical Society, 1986.

Zirkle, Conway. "Benjamin Franklin, Thomas Malthus and the United States Census." In Hindle, Brooke, editor, *The History of Science Selections from ISIS*. New York: Science History Publications, 1976.

NOTES

INTRODUCTION

xi *One day in April 1755:* B.F. to Peter Collinson, Aug. 25, 1755, *B.F. Papers,* Volume 6, pp. 167–168. Van Doren, *Benjamin Franklin,* pp. 180–182. The Colonel Tasker referred to is either Benjamin Tasker, mayor of Annapolis, or his son, also named Benjamin Tasker, a delegate to the Albany Congress.

xii *enters and leaves:* Breitwieser, *Cotton Mather and Benjamin Franklin,* pp. 208–209.

xii *Nature alone met him:* Becker, *Benjamin Franklin,* p. 36.

xii *inadvertently took:* B.F. letter (possibly to John Franklin), December 25, 1750, *B.F. Papers,* Volume 4, p. 82.

xiii *We learn by chess:* Benjamin Franklin. "The Morals of Chess." *B.F. Papers,* Volume 29, p. 750.

xiii *Thunderstorms "inflam'd my Curiosity":* B.F. to Peter Collinson, September 1753. *B.F. Papers,* Volume 5, pp. 69–70.

xiv *[On] Tuesday last: Pennsylvania Gazette* Extracts 1732. *B.F. Papers,* Volume 1, pp. 272–275.

xvi *The usual portrayal of Franklin:* I. Bernard Cohen, *Benjamin Franklin: Scientist and Statesman,* p. 474.

CHAPTER ONE

3 *great Excellence lay:* Benjamin Franklin, *The Autobiography of Benjamin Franklin,* pp. 54–55.

3 *I do not remember when I could not read: Autobiography,* p. 53.

4 *If the Buds are so precious: Literary Diary of Ezra Stiles,* Volume II, pp. 375–376; quoted in Aldridge, *Benjamin Franklin and Nature's God,* 1967.

4 *it would be a vast saving:* William Temple Franklin, *Memoirs,* Volume 1, p. 447.

4 *Living near the water: Benjamin Franklin, Autobiography,* pp. 53–54; Van Doren, *Benjamin Franklin,* p. 55.

4 *Having then engaged:* Quoted in Tourtellot, *Benjamin Franklin,* p. 161.

5 *Under Apprehensions: Autobiography,* p. 57.

5 *In a little time I made: Autobiography,* p. 59.

5 *perhaps gave me a Turn of Thinking: Autobiography,* p. 58.

5 *By the time he was a teenager:* Thomas, *Religion and the Decline of Magic,* p. 81.

6 *The new [Newtonian] mechanics:* Struik, *Yankee Science in the Making,* p. 56.

6 *As Voltaire, who helped:* Voltaire, *Oeuvres,* Volume XXII, p. 130; in Becker, *The Heavenly City of the Eighteenth-Century Philosophers,* p. 60.

7 *an enchanted universe:* Hall, *Worlds of Wonder, Days of Judgment,* p. 71.

7 *In place of unacceptable moral chaos:* Thomas, p. 107.

7 *Without doubt the Lord:* Quoted in Mather, *Magnalia Christi Americana,* p. 5.

8 *Public unease over the executions:* Taylor, *American Colonies,* p. 185.

9 *Thus an entire generation had grown up:* Smallpox, or "small-pocks," first known in third century A.D. China, returned to northern Europe with the homecoming Crusaders of the eleventh through thirteenth centuries, and was introduced to southern Europe by the Moors. By the seventeenth century it was endemic in both Europe (Queen Mary died of it in 1694) and colonial America.

9 *have stirred up the Anger:* Quoted in Perry Miller, *The New England Mind,* p. 346.

11 *For my own part:* Cotton Mather to Dr. John Woodward, July 12, 1716; G. L. Kittredge, "Lost Works of Cotton Mather," *Mass. Hist. Soc. Proc.,* Volume XLV, 1911–1912, p. 422; quoted in Kraus, *The Atlantic Civilization,* pp. 207–208.

12 *Every part of matter is peopled:* Mather, *The Angel of Bethesda,* p. 43.

12 *He also had little patience:* William Douglass. *A Summary . . . of the First Planting, Progressive Improvements and Present State of the British Settlements in North America,* 2 volumes. Boston: Daniel Fowle, 1751, Volume 2, p. 6; quoted in Maxine Van de Wetering, "A Reconsideration of the Inoculation Controversy."

13 *No one need fear: Boston Gazette,* July 17, 1721.

13 *negroish evidence:* Silverman, *Selected Letters of Cotton Mather,* p. 340. Cotton Mather, *Some Account of what is said of Inoculating or Transplanting the Small Pox by the learned Dr. Emanuel Timonius and Jacobus Pylarinus,* Boston, 1721; quoted in Kraus, pp. 209–210.

13 *Men of Piety and Learning: Boston Gazette,* July 31, 1721.

14 *The public, increasingly:* Public trust in the clergy was already at a low ebb. The previous year the ministers had meddled in a local financial crisis, admonishing residents that the situation would improve if they practiced virtue and stopped drinking rum. But when the crisis worsened,

Bostonians sensed that the churchmen were useless at solving real-world problems. Warden, *Boston*, pp. 84–85.

14 *a vigorous little sheet:* Emery, *The Press and America*, p. 38.

14 *a Notorious, Scandalous:* W. C. Ford, "Franklin's *New England Courant*," *Mass. Historical Society Proceedings*, Volume 57, April 1924; in Lopez and Herbert, *The Private Franklin*, pp. 11–12.

15 *I cannot but pity poor Franklin:* Quoted in Isaacson, *Benjamin Franklin*, p. 23.

15 *sarcastically proposing that:* New-England Courant, August 14, 1721.

15 *Increase stepped in:* Perry Miller, p. 360.

15 *Warnings are to be given:* Cotton Mather, Diary, 1709–1724, in *Collections of the Massachusetts Historical Society*, no. 8, seventh series, Boston, 1912, p. 663, quoted in Leonard, *The Power of the Press*, p. 23.

15 *James, in the next issue:* New-England Courant, December 25, 1721. The appearance of newspapers had a profound influence on New England society. Among other effects, by regularly publishing information about lost articles, missing cows, and runaway apprentices, they dispelled the Puritans' need for magic charms and amulets to reckon with inexplicable events.

16 *Cotton Mather, You dog:* Anti-inoculationist sentiment hardly diminished in the years to come. As late as 1773, a mob burned a hospital at Marblehead for giving inoculations, and tarred and feathered four men believed to have smallpox.

16 *Mather* was *proven right:* Boston smallpox data in Tourtellot, p. 260.

16 *By introducing into the covenant theory:* Perry Miller, pp. 363–364.

17 *Surely parents will no longer refuse:* B.F.'s Preface to Dr. Wm. Heberden's "Some Accounts of the Success of Inoculation for the Small-Pox," London, Feb. 16, 1759; in *B.F. Papers*, Volume 8, pp. 281–286. The battle against smallpox would take a great stride forward in the late 1790s with the introduction of vaccination by the English physician Edward Jenner. Investigating the folk wisdom that dairy maids who had contracted cowpox did not get smallpox, Jenner pioneered the method of thwarting one malady (smallpox) with pustules from another (cowpox). After the publication in 1798 of his *Inquiry into the Causes and Effects of the Variolae Vaccinae*, vaccination became the accepted means of deterring the disease.

17 *When I was a boy:* B.F. to Samuel Mather, May 12, 1784; in Benjamin Franklin, *The Writings of Benjamin Franklin*, Volume 9, pp. 208–210. The full title of the book, published in 1710, was *Bonifacius, Essays to Do Good*. Extremely popular, it went through eighteen printings. In *Essays*, Mather suggested that people write down resolutions and revisit them often to gauge their progress, an idea later put into practice by Franklin.

18 The Spectator *introduced an:* The Spectator *was published daily from March 11, 1711, to December 6, 1712.

19 *they contented themselves:* Autobiography, p. 69.

19 *During my Brother's Confinement:* Autobiography, p. 69.

20 *I had already made myself:* Autobiography, p. 71.

21 *So I sold some of my books:* Autobiography, p. 71.

21 *which I took extremely amiss:* Autobiography, p. 68.

21 *I fancy his harsh and tyrannical:* Autobiography, p. 69.

21 *The rebel in him:* Lopez and Herbert, p. 11.

21 *He received me in his library:* B.F. to Samuel Mather, May 12, 1784; in *Writings*, Volume 9, pp. 208–210. The Mather-Franklin connection deservedly intrigues many historians. Mather was a highly literate Puritan cleric sensitive to the fresh currents of the arriving Enlightenment; Franklin's "main meaning for the American psyche," according to the novelist John Updike, was "a release into the Enlightenment of the energies cramped under Puritanism." See John Updike, "Many Bens," *The New Yorker*, February 22, 1988.

CHAPTER TWO

23 *William Penn's city:* Because of Penn's novel approach toward the Indians—he acknowledged their right to own land—the colony actually attracted Native Americans from distant places where they had been abused. These refugees settled on Philadelphia's western frontier, thus providing a cushion between the city and any potentially hostile tribes to the west.

24 *No colonial city:* Bridenbaugh, *Rebels and Gentlemen*, p. 27, p. 28, p. 371.

24 *showed off his swimming skills:* Franklin's exhibitions of swimming prowess were well-known. Once, on a Sunday outing on the Thames with friends, he dove overboard and swam alongside the boat for three and a half miles, "from near Chelsea to Blackfryars, performing on the Way many Feats of Activity both upon and under Water, that surpriz'd and pleas'd those to whom they were Novelties" (Benjamin Franklin, *The Autobiography of Benjamin Franklin*, p. 104). Sir William Wyndham, an aristocrat who learned of Franklin's dip in the Thames, inquired whether he might teach his sons to swim, causing Franklin to briefly entertain the idea of traveling through Europe giving aquatic lessons to the gentry.

25 *I have never fixed:* Quoted in Van Doren, *Benjamin Franklin*, p. 69.

26 *Truth and sincerity:* Benjamin Franklin, "Journal of a Voyage." *B.F. Papers*, Volume 1, pp. 72–99. See also Phyllis Franklin, *Show Thyself a Man*, p. 61.

27 *one of his first attempts at a literary hoax:* Like his earlier interest in the technique of Socratic inquiry, Franklin's love of hoaxing seems to have been born of his desire to develop alternative and effective means of persuasion that did not rely upon traditional argumentation. His best-known hoax was probably "The Speech of Polly Baker," in which a destitute woman brought before a court defends eloquently the fact that she has had a number of children out of wedlock, in the process so impressing one of the judges that he proposes marriage. Written and published by Franklin in the 1740s, the piece was still being occasionally reprinted as genuine a century later.

27 *conceited scribbler:* Poor Richard's Almanac, 1733, 1734; Titan Leeds's *American Almanack*, 1734; quoted in Isaacson, *Benjamin Franklin*, p. 96.

27 *they were great things to me:* Autobiography, p. 197.

28 *All the Inhabitants: B.F. Papers*, Volume 7, p. 317.

28 *In order to secure: Autobiography*, pp. 125–126.

28 *The industry of that Franklin: Autobiography*, p. 119.

29 *Our Benjamin Franklin: B.F. Papers*, Volume 3, p. 458n.

30 *An interest in science:* Lemay, "Benjamin Franklin: Universal Genius, the Renaissance Man in the Eighteenth Century," p. 14.

30 *an ever expanding web:* These natural philosophy connections were multi-generational. Colden's son David pursued and corresponded with Franklin about electrical experiments; his daughter Jane learned the Linnaean system of plant classification and became a capable botanist. Her lavishly illustrated *Flora of New York* wound up in the British Museum. *Travels through North and South Carolina, Georgia, E. & W. Florida, etc.*, by Bartram's artist son William, whose nature drawings anticipated James Audubon's, remains one of the most captivating books ever written about America's wilderness and its native Indians, and directly inspired poets and writers of the Romantic era such as Keats, Coleridge, and Chateaubriand.

31 *The closer we get:* Quoted in "The Two Faces of Linnaeus" by Sten Lindroth; in Frangsmyr, editor, *Linnaeus*, p. 3, p. 16.

32 *We repair our bodies:* Kraus, *The Atlantic Civilization*, p. 201.

32 *The major gardens:* Gardens reflected the belief that nature could be classified and made symmetrical, and glorified the powerful reach of affluent societies in bringing commodities and exotica from afar. In order to suggest an Edenic world of beauty and treachery, many large eighteenth-century estate gardens contained elaborate mazes, sound effects like thunder or volcanoes, bizarre topiary, and even trick water jets that sprayed unsuspecting visitors.

33 *A knowledge of science:* Bridenbaugh, p. 314. One of Collinson's role models was surely Sir Hans Sloane, who gathered specimens so vigorously that, at his death in 1753 at the age of ninety-three, his collections of books, plants and animals, and coins and antiquities became the basis for the British Museum. (He had realized the need to better safeguard his collection, it was said, when the composer George Frideric Handel visited one morning and, mid-conversation, set a buttered muffin on top of a rare manuscript.) Sloane is credited with inventing the drink known as hot chocolate.

33 *Keep the chain:* Bartram to B.F., July 29, 1757; *B.F. Papers*, Volume 7, pp. 246–247.

34 *very great Forests:* Bartram to Collinson, April 26, 1737; quoted in Stearns, *Science in the British Colonies of America*, p. 582.

34 *greatest natural botanist:* Bartram's book was *Observations on the Inhabitants, Climate, Soil, Rivers, Productions, Animals, and Other Matters Worthy of Note, Made by Mr. John Bartram in his travels from Pennsilvania to Onondaga, Oswego, and Lake Ontario in Canada*. It was published in London in 1751.

35 *Rattlesnakes seem: Pennsylvania Gazette*, May 9, 1751.

35 *Franklin's wide acquaintance:* Increase Mather and his sons had, in 1683, started a Boston Philosophical Society, but the group was never formally

organized and seems to have disbanded about 1688. Cadwallader Colden had suggested the formation of an intercolonial scientific entity as far back as 1728, and in 1730 there reportedly gathered in Newport, Rhode Island, a "Society for the Promotion of Knowledge and Virtue by a Free Conversation."

35 *All new-discovered Plants:* "A Proposal for Promoting Useful Knowledge," in *B.F. Papers,* Volume 2, pp. 378–383.

36 *air was drawn in:* "An Account of the New Invented Pennsylvania Fire-Places," *B.F. Papers,* Volume 2, pp. 419–446. Another valuable feature of the stove was that it was slow-burning and thus required less wood.

37 *As he would with all his inventions:* The first popular electrical innovation to follow the lightning rod, the electromagnetic telegraph, created a bitter three-way dispute among its inventor, Samuel Morse, and two men whose efforts or ideas Morse had built upon, Joseph Henry, head of the Smithsonian Institute, and Charles Thomas Jackson.

38 *They were imperfectly performed: Autobiography,* pp. 140–141.

38 *Spencer offered to distribute:* Franklin and the Library Company had performed a similar service at least once before, arranging in 1740 to host lectures in Philadelphia by Isaac Greenwood. One of the first academic scientists in America, Greenwood fell tragically short of his potential. In 1738 he was dismissed from Harvard, where he held the distinguished Hollis Chair, for intemperance, and died while on a voyage to South Carolina in 1745.

38 *The Animalculae:* Broadside published by Franklin. Reproduced in I. Bernard Cohen, *Benjamin Franklin's Science,* p. 59.

39 *most judicious and experienced:* Bridenbaugh, p. 269.

39 *A little Cool Air:* Quoted in I. Bernard Cohen, *Benjamin Franklin's Science,* p. 53.

39 *put them into very brisk:* William Smith manuscript, Volume 5, p. 254; quoted in I. Bernard Cohen, *Benjamin Franklin's Science,* p. 46. Spencer continued to lecture along the Atlantic seaboard until 1751, when he took up residence in Anne Arundel County, Maryland. He died there in 1760.

40 *In the* Opticks: I. Bernard Cohen, *Science and the Founding Fathers,* p. 141.

40 *As in mathematics:* Newton, *Opticks.*

41 *with superstitious reverence:* Benjamin, *The Age of Electricity,* pp. 2–3.

41 *First-century philosopher:* The subject of electricity enters the sphere of scientific inquiry in the sixth century B.C. with Thales of Miletus, who mentioned amber's qualities. Three hundred years later, Theophrastus, one of Aristotle's pupils, described it in a treatise on gems.

42 *Gilbert may justly be called:* Priestley, *The History and Present State of Electricity,* p. 5. One reason Gilbert's electrical musings may have gone ignored was that Francis Bacon, in his seminal *Novum Organum* (1620), dismissed Gilbert's methods and conclusions about electricity as "fables." Thus, Bacon, while exercising so liberating an effect on science by promoting the concept of inductive as opposed to a priori reasoning, may have inadvertently squelched inquiry into one of the earth's greatest unknown natural forces. Benjamin, pp. 7–8.

43 *if an insulated charged:* D. C. Miller, *Sparks, Lightning, Cosmic Rays,* p. 14.

45 *new but terrible experiment:* Musschenbroek to Réaumur, January 20, 1746; quoted in Heilbron, *Electricity in the 17th and 18th Centuries*, p. 313. The Leyden jar could hold its charge for hours or even days. Patricia Fara reports a Leyden battery now in a museum in Europe that has "an adjustable dial with four settings: detonating cannon, altering a compass needle, killing small animals, and melting wire." Fara, *An Entertainment for Angels*, pp. 56–57. Franklin was not the only person in the colonies to pursue "electrics." John Winthrop at Harvard, William Claggett (a New England clock-maker), and Cadwallader Colden in the Hudson River Valley were among those who tried to duplicate the demonstrations of Gray and Du Fay.

45 *how far an electrical shock could be:* The ability of electricity to instantaneously travel substantial distances would be one of the founding principles of the electromagnetic telegraph, invented by Samuel Morse in the 1830s.

45 *He suspected:* I. Bernard Cohen, *Franklin and Newton*, p. 435.

46 *Written by Albrecht von Haller:* Haller discussed the recent work by the Germans Georg Mathias Bose, Christian Hausen, and Johann Winkler, who had improved upon existing methods of generating large charges of electricity.

46 *Electricity is a vast country: The Gentleman's Magazine*, Volume 15, 1745, pp. 193–197; quoted in I. Bernard Cohen, *Benjamin Franklin's Science*, p. 63.

46 *The first Drudgery:* "A Proposal for Promoting Useful Knowledge," May 14, 1743. *B.F. Papers*, Volume 2, pp. 378–383.

47 *I am in a fair Way:* B.F. to Colden, September 29, 1748, *B.F. Papers*, Volume 3, pp. 317–320.

47 *I never was before engaged:* B.F. to Collinson, March 28, 1747, *B.F. Papers*, Volume 3, p. 115.

48 *we had at length: Autobiography*, pp. 240–241.

49 *first primitive views:* I. Bernard Cohen, *Franklin and Newton*, p. 309. My description of Nollet's and Franklin's electrical theories is indebted to Roderick Home, professor of the history of science at the University of Melbourne.

49 *Perhaps reflecting:* Fara, p. 114.

50 *"invisible" particles:* Seeger, *Benjamin Franklin*, p. 13. Thomson, working with cathode rays, found that subatomic particles deflected by magnetic and electrical fields were smaller than the smallest known atom—about 1/1,000 the size of the hydrogen atom—and appeared to be a component of all matter. For identifying electrons—Franklin's "particles extremely subtile"—Thomson was awarded the Nobel Prize in 1906.

50 *any excess that they receive:* I. Bernard Cohen, *Franklin and Newton*, p. 303.

50 *Franklin's colleague Thomas Hopkinson:* Thomas Hopkinson died in 1751, but his son Francis, later a signer of the Declaration of Independence, pursued electrical studies. Franklin, while posted in Paris, authorized Francis to use the electrical apparatus in his own house in Philadelphia.

50 *If you present the point:* B.F. to Collinson, May 25, 1747, *B.F. Papers*, Volume 3, p. 127.

51 *Kinnersley, grew so adept:* Kinnersley lecture broadside dated April 25, 1753,

in I. Bernard Cohen, *Benjamin Franklin's Science*, Appendix 1. Kinnersley had designed the so-called "Magic Picture" demonstration, in which the gilt in a picture frame was electrified. Early on, Franklin and Kinnersley considered this a means of honoring the king, although later, when sentiments between the colonies and England soured, a way to mock him. One of Kinnersley's bravura performances was his electric "burning in effigy" of the British tormentor of Franklin, Alexander Wedderburn. With Franklin's help, Kinnersley secured a teaching job at the Academy, later the University of Pennsylvania.

52 *that he had lived eight hundred miles upon:* B.F. to John Lining, March 18, 1755; *B.F. Papers,* Volume 5, pp. 521–527.

52 *On some further experiments:* B.F. to Collinson, August 14, 1747, *B.F. Papers,* Volume 3, p. 171.

54 *implore the Blessing: Autobiography,* pp. 184–185.

54 *the principal Mover:* James Logan to Thomas Penn, November 24, 1749; in *B.F. Papers,* Volume 3, p. 185.

55 *The naming was a playful allusion:* Fara, p. 26.

55 *To put an end to them:* B.F. to Collinson, April 29, 1749, *B.F. Papers,* Volume 3, pp. 352–365.

56 *Electrical fluid agrees with lightning:* B.F. to John Lining, March 18, 1755, *B.F. Papers,* Volume 5, pp. 521–527.

57 *the great operations of nature:* Benjamin Franklin, "Of Lightning and the Method (Now Used in America) of Securing Buildings and Persons from Its Mischievous Effects." *B.F. Papers,* Volume 14, pp. 260–264.

57 Let the experiment be made: In scientific minutes of his experiments made by B.F. on November 7, 1749, later sent to John Lining on March 18, 1755, *B.F. Papers,* Volume 5, pp. 521–527.

58 *I have seen myself:* Pliny the Elder, *Natural History,* pp. 34–35.

59 *a good wire communication:* Schonland, "The Work of Benjamin Franklin on Thunderstorms."

59 *May not the knowledge:* B.F. to Collinson, "Opinions and Conjectures," July 29, 1750, *B.F. Papers,* Volume 4, pp. 9–34; quoted in I. Bernard Cohen, *Benjamin Franklin's Experiments,* p. 222.

60 *Clear Intelligent Stile:* F. Christensen, "John Wilkins and the Royal Society Reform of Prose Style," *Modern Language Quarterly,* Volume VII, 1946, pp. 279–291; quoted in Kraus, *The Atlantic Civilization,* p. 6.

61 *Informed that Godfrey:* Bridenbaugh, pp. 307–308.

CHAPTER THREE

64 *Thor's handiwork:* In Europe and the British Isles, the term "thunderstones" was given to mysterious objects occasionally unearthed by farmers that, it was believed, fell to earth with the rain: many were large fossils or relics of prehistoric animals; others may have been small meteorites. Lightning actually does at times leave behind some curious stone-like forms—hollow quartz-like cylinders known as *fulgurites* that are fused instantly when lightning strikes sand.

64 *Like as out of a burning:* Pliny the Elder, *Natural History,* p. 33.

65 *would not hurt anything:* Pliny the Elder, p. 36.

65 *We hear from Virginia:* Pennsylvania *Gazette* Extracts, 1736, in *B.F. Papers,* Volume 2, p. 160.

65 *Such stories:* "Lightning-picture" anecdotes are from *Appleton's Journal of Popular Literature, Science, and Art,* October 1869.

66 *It was, indeed, no great step:* White, *History of the Warfare of Science,* p. 337.

67 *there was scarce a great abbey:* Tomlinson, *The Thunder-Storm,* p. 180.

68 *Whensoever this bell:* Quoted in White, pp. 346–347.

68 *I, a priest of Christ:* Ibid., p. 340.

68 *The fetish's power:* Ibid., p. 343.

68 A Proof That the Ringing of Bells: Arago, *Meteorological Essays,* p. 182.

69 *the well-known legal concept of "an act of God":* Thomas, *Religion and the Decline of Magic,* p. 84.

69 *The surest remedy:* Quoted in White, p. 350.

69 *Before, I used to be:* Edwards, *Works,* Volume I, p. 62.

70 *much amused:* Quoted in I. Bernard Cohen, *Franklin and Newton,* p. 288.

70 *the electric fire:* D. C. Miller, *Sparks, Lightning, Cosmic Rays,* pp. 57–58.

70 *If someone:* Nollet, *Leçons de Physique Expérimentale,* Volume IV, p. 64; quoted in Still, *The Soul of Amber,* p. 129.

71 *much disposed to like the World:* Van Doren, *Benjamin Franklin,* pp. 124–126.

71 *had little more faith:* Aldridge, *Benjamin Franklin and Nature's God,* p. 8.

71 *If men are so wicked: B.F. Papers,* Volume 7, pp. 293–295.

71 *Here will I hold:* Joseph Addison, *Cato, A Tragedy,* pp. 15–18; quoted in *B.F. Papers,* Volume 1, p. 101.

72 *a talent for happiness:* Stifler, *The Religion of Benjamin Franklin,* p. 111.

72 *I thought of your excessively strict:* B.F. to Jared Ingersoll, December 11, 1762, in *B.F. Papers,* Volume 10, pp. 174–176.

73 *To Franklin, any and all:* For implying that God is a human construct, Shaftesbury's theories were deemed heretical, and some of the French *philosophes* he influenced—including Denis Diderot and the learned men of the *Encyclopédie*—spent time in prison or exile for their views. Furbank, *Diderot,* pp. 25–26.

73 *inspire, promote or confirm:* Benjamin Franklin, *The Autobiography of Benjamin Franklin,* p. 146.

73 *Because Franklin thought God wanted:* Benjamin Franklin, "Articles of Belief and Acts of Religion," November 20, 1728, in *B.F. Papers,* Volume 1, pp. 101–109.

74 *at the last Day:* B.F. to Josiah and Abiah Franklin, 1738; quoted in Phyllis Franklin, *Show Thyself a Man,* pp. 64–65.

74 *making long Prayers:* B.F. to Joseph Huey, June 6, 1753; *B.F. Papers,* Volume 4, pp. 503–506.

75 *multitudes were seriously:* Ezra Stiles, quoted in Taylor, *American Colonies,* p. 352.

75 *Free choice . . . had radical:* Taylor, p. 354.

75 *perhaps the most charismatic:* Brands, *The First American,* p. 138.

75 *pleasant hunting:* Luke Tyerman, *The Life of the Rev. George Whitefield,* London, 1877, pp. 332–333; quoted in *B.F. Papers,* Volume 5, p. 429.

75 *He was at first permitted: Autobiography,* p. 179.

76 *I happened . . . to attend:* Autobiography, p. 177.

77 *The two men:* Stout, *The Divine Dramatist*, p. 222.

78 *For Franklin personally:* A counterpoint to Franklin's cautious approach was that of his friend Cadwallader Colden, who in about 1743 began a questionable investigation into the cause of gravity. Seeking answers to questions that had stumped every natural philosopher of his age, including Newton, Colden did not adequately study other works in the field or thoroughly grasp Newton's ideas on inertia and the laws of motion. His treatise, *An Explication of the First Causes of Action in Matter; and the Cause of Gravitation*, published in 1746, struck even his friends as incomprehensible. The failure was made even more embarrassing by Colden's large ambitions for the work.

78 *an engine of research:* Furbank, p. 37.

79 *regarded civil society:* Furbank, p. 23.

79 *the most important scientific contribution:* Hindle, *The Pursuit of Science in Revolutionary America*, p. 77.

79 *conducts us by a train of facts:* John Fothergill, in his Preface to *Experiments and Observations, Made at Philadelphia*, 1751.

79 *we Americans are not half-removed:* Quoted in Ellis, *After the Revolution*, p. 11.

81 *the honest ecclesiastic:* Priestley, *The History and Present State of Electricity*.

82 *In following the path:* Quoted in I. Bernard Cohen, *Benjamin Franklin's Science*, p. 73.

82 *The effect on the public mind:* Schonland, *The Flight of Thunderbolts*, p. 21.

82 *the greatest discovery that has been made: Gentleman's Magazine*, June 1752.

83 *toy science:* I. Bernard Cohen, *Franklin and Newton*, pp. 286–287.

83 *almost inaccessible:* Watson, quoted in I. Bernard Cohen, *Franklin and Newton*, p. 490.

83 *a subject of serene contemplation:* Harris, *Observations of the Effects of Lightning on Floating Bodies*, p. 3.

83 *He had in fact developed:* Several others verified Franklin's kite experiment, including Ebenezer Kinnersley and John Lining. Peter Van Musschenbroek, one of the inventors of the Leyden jar, was inspired by Franklin's example to use a kite for more general electric research. Giambatista Beccaria and Tiberius Cavallo also used kites. A. Lawrence Rotch, the nineteenth-century American physicist, and aviation pioneers Wilbur and Orville Wright were prominent among those who adapted the kite for use as a research tool. Historians of electricity are familiar also with the claims of Jacques de Romas of Bordeaux and Jean M. Mazeas, a Fellow of the Royal Society, both of whom said they either flew or conceived of the kite experiment before Franklin. Mazeas, if he did conduct such an experiment, appears to have kept it a secret until after Franklin had come forward. Romas admitted he had not flown his kite until May 14, 1753, although he insisted he had *thought of it* prior to June 1752. Lacking documentation to support this fact, Romas was disbelieved by Franklin's supporters, although Franklin himself was reported by Montesquieu to have toasted Romas's health one night at a dinner in Paris. Riskin, *Science in the Age of Sensibility*, p. 101.

84 *A heavy cloud above the city:* Richards, *Electron, or Pranks of the Modern Puck.*

85 *While the rest of Philadelphia:* Westcott, *A History of Philadelphia*, p. 420.

86 *electric fire very copiously:* Priestley, *The History and Present State of Electricity*, p. 217.

87 *At the moment:* White, p. 364.

87 *We hear from Susquehanna: Pennsylvania Gazette*, August 13, 1752.

87 *a Bull and two cows: Pennsylvania Gazette*, July 12, 1753.

88 *It was very remarkable: Pennsylvania Gazette*, August 5, 1752.

88 *the fluid passes in the walls:* Benjamin Franklin, "Of Lightning, and the Method Now Used in America of Securing Buildings and Persons from Its Mischievious Effects," 1769. *B.F. Papers*, Volume 14, p. 262.

88 *will go considerably out of a direct Course:* B.F. to Collinson, September 1753. *B.F. Papers*, Volume 5, pp. 68–79.

89 *In September:* Quoted in I. Bernard Cohen, *Benjamin Franklin's Science*, p. 89.

89 *I was one night waked:* Benjamin Franklin, "Experiments Supporting the Use of Pointed Lightning Rods." August 18, 1772. *B.F. Papers*, Volume 19, pp. 244–255; quoted in I. Bernard Cohen, *Benjamin Franklin's Science*, p. 90.

89 *In summer 1752:* I. Bernard Cohen, *Benjamin Franklin's Science*, p. 91.

89 *making a Machine or Kite: Philosophical Transactions*, Volume 47, 1751–52, pp. 565–567.

91 *It has pleased God: Poor Richard's Almanac*, 1753, published October 1752.

91 *One of the first fully:* Ebenezer Kinnersley to B.F., March 12, 1761; *B.F. Papers*, Volume 9, pp. 291–293. The Royal Society awarded Franklin its coveted Copley Medal in 1753. "Though some others might have begun to entertain suspicions of an analogy between the effects of lightning and electricity, yet I take Mr. Franklin to be the first who, among other curious discoveries, undertook to shew from experiments, that the former owed its origin entirely to the latter," observed the Earl of Macclesfield in awarding the medal to Franklin. "Electricity is a neglected subject, which not many years since was thought to be of little importance . . . nor was anything worth much notice expected to ensue from it." But now, declared the Earl, it "appears to have a most surprising share of power in nature." See Lokken, *Meet Dr. Franklin*, 1998.

93 *vulgar errors, extravagant fears:* Nollet, *Programme ou Idée Générale d'un Cours de Physique Expérimentale*, pp. xviii–xix.

93 *way of reasoning:* Lelarge de Lignac, *Lettres*, 1751, p. 1; quoted in Heilbron, p. 346.

94 *the King's Thanks:* Quoted in I. Bernard Cohen, *Benjamin Franklin's Science*, pp. 74–75.

94 *is dying of chagrin:* Buffon, *Correspondance de Buffon de 1729 à 1788*, p. 84.

94 *When my papers were first published:* B.F. to Ingenhousz, June 21, 1782, *B.F. Papers*, Volume 35, pp. 549–551; quoted in I. Bernard Cohen, *Benjamin Franklin's Science*, p. 105.

95 *the Sea might possibly be:* B.F. to James Bowdoin, Jan. 24, 1752, *B.F. Papers*, Volume 4, pp. 256–259.

95 *Sea Water in a Bottle:* B.F. to Collinson, September 1753, *B.F. Papers,*
Volume 5, pp. 68–79.

95 *Perhaps some future experiments:* These "future experiments" performed
by Franklin's scientific descendants suggest that the earth's atmosphere
functions as a kind of giant hydroelectric machine. It is thought that
swift updrafts of air tear tiny water droplets in the lower atmosphere
apart. The larger ones fall toward earth as rain, and the smaller ones rise
upward where they become charged with negative electricity. Since the
ice crystals that form near the tops of thunderclouds are charged positively, the cloud is charged with opposing electrical forces. The electrical discharge of lightning is produced as the different charges attempt
to neutralize one another or achieve electrical equilibrium with the
earth. Lightning researchers, however, concur that much remains unknown about the process by which the electrification of the atmosphere
occurs.

96 *as impious to ward off:* Benjamin, *The Age of Electricity,* p. 26.

96 *All these iron points:* Schonland, p. 31.

96 *my business to enquire:* Quoted in Heilbron, *Electricity in the 17th and 18th
Centuries,* p. 391.

97 *globe of blue fire:* Priestley, p. 311.

97 *They opened a vein:* Priestley, pp. 331–332.

98 *an enviable one:* B.F. *Papers,* footnote, Volume 5, p. 155.

98 *the new Doctrine of Lightning:* "Account of the Death of Georg Richmann,"
in *Pennsylvania Gazette,* March 5, 1754; *B.F. Papers,* Volume 5, pp. 219–221;
see also "An Account of the Death of Mr. George William Richmann" in
Philosophical Transactions, Volume 49, 1755–56.

99 *farr from Dealing candidly:* Peter Collinson to B.F., July 20, 1753; *B.F.
Papers,* Volume 5, pp. 12–14.

99 *Surely the Thunder:* B.F. to Colden, April 12, 1753; *Letters and Papers of
Cadwallader Colden,* Volume 4, 1748–1754, Collections of the New York
Historical Society (1920), p. 382; quoted in I. Bernard Cohen, *Benjamin
Franklin's Science,* p. 141.

99 *'tis the earth that strikes into the clouds:* B.F. to Collinson, September
1753, *B.F. Papers,* Volume 5, pp. 68–79. See also B.F. to Colden, April 12,
1753, *B.F. Papers,* Volume 4, pp. 463–465.

100 *his was the first hypothesis:* I. Bernard Cohen, *Benjamin Franklin's Science,*
pp. 141–142.

100 *That it is not always directed:* From "Junto Minute Book in the APS Library,
Junto Society Meeting of February 15, 1760," quoted in I. Bernard
Cohen, *Benjamin Franklin's Science,* pp. 142–143.

100 *It is perhaps not so extraordinary:* B.F. to Winthrop, July 2, 1768, *B.F. Papers,*
Volume 15, pp. 166–172.

CHAPTER FOUR

102 *Our fears came upon us: Boston Gazette,* December 1, 1755.

103 *God, in his holy Providence: Boston Evening Post,* November 24, 1755.

103 *'Tis* Sin, *and that* only: *Boston Gazette,* November 24, 1755.

103 *Some favored the idea:* Thomas, *Religion and the Decline of Magic,* pp. 80–81.

103 *A series of additional tremors:* Houston, "When the Great Earthquake Struck New England." Scientists estimate that the New England earthquake of 1755 was probably a 6.0 on the Richter scale; by comparison, the devastating 1906 San Francisco earthquake was an 8.3.

103 *The crew aboard a ship: Pennsylvania Gazette,* January 22, 1756; quoted in *B.F. Papers,* Volume 6, p. 404.

103 *The catastrophe thus stymied:* Modern historians believe that the actual death toll of the Lisbon earthquake was probably between ten thousand and fifteen thousand.

104 *Lisbon . . . was the Enlightenment's "Auschwitz":* Neiman, *Evil in Modern Thought,* p. 1, pp. 240–250.

104 *No Doubt natural Causes:* Houston, "When the Great Earthquake Struck New England," p. 104.

105 *The more Points of Iron:* Thomas Prince, *Earthquakes the Works of God and Tokens of His Just Displeasure . . . on Occasion of the late Dreadful Earthquake which happened on the 18th of Nov. 1755.*

106 *a kind of universal fire: Pennsylvania Gazette,* Dec. 8–22, 1737; see Aldridge, "Benjamin Franklin and Jonathan Edwards on Lightning and Earthquakes."

106 *produc'd by a thousand Miles:* Stukeley, "Letter to the President on the Causes of Earthquakes," *Philosophical Transactions,* Volume 46, Appendix, 1750.

106 *Noting that major earthquakes often occurred:* Stukeley, "Letter to the President on the Causes of Earthquakes," *Philosophical Transactions.*

106 *"chastening rod":* Ibid.

107 *two Specimens:* Elizabeth Hubbart to B.F., Feb. 16, 1756, *B.F. Papers,* Volume 6, p. 404.

107 *Winthrop personified:* Winthrop was the second Hollis Professor at Harvard; the first, Isaac Greenwood, occupied the chair from 1728 to 1738.

107 *in its highest perfection:* Winthrop's "Commonplace Book," 1728–1732; quoted in Michael N. Shute, "John Winthrop: Professional Science, Imagination, and Early American Culture," in Shute, *The Scientific Work of John Winthrop.*

107 *Since 1746:* Known to jot down notations of current meteorological conditions three times a day, Winthrop was one of the first Americans (along with Thomas Jefferson) to make astronomical observations. As early as 1740 he reported to the Royal Society, on a transit of Mercury one of fourteen papers he would submit en route to becoming a Fellow.

108 *our buildings were rocked:* Winthrop, "An Account of the Earthquake Felt in New England," pp. 1–18.

109 *It is as much our duty:* Hindle, *The Pursuit of Science in Revolutionary America,* pp. 95–96.

109 *He conceded there was an old folk belief:* The sulfurous smell was actually ozone, an unstable form of oxygen. At the time, neither ozone nor oxygen had been discovered.

109 *I cannot but esteem it: Boston Gazette,* January 28, 1756; quoted in Eleanor Tilton, "Lightning Rods and the Earthquake of 1755."

110 *discourage the use of the iron-points:* Winthrop, Appendix to a Lecture on Earthquakes: "Concerning the Operation of Electrical Substance in Earthquakes; and the Effects of Iron Points."

110 *[I] cannot believe:* Winthrop, "Appendix to a Lecture on Earthquakes."

110 *The most pious believer:* A final blow to New England's reliance on providence was right around the corner, for in 1758 Halley's comet reappeared on schedule. This event, predicted decades earlier by Newton and Edmond Halley, helped free comets once and for all from the realm of enchantment in which they had long been held. The comet, which reappears about every 76 years, visited most recently in 1986.

111 *One person who took a great interest:* Haraszti, "Young John Adams on Franklin's Iron Points."

112 *be always in the heads of the Wild:* Penn to Richard Peters, June 9, 1748, *Penn Letter Book,* Volume II, p. 232, Historical Society of Pennsylvania; quoted in *B.F. Papers,* Volume 3, p. 186.

112 *For my own part:* William Strahan to Deborah Franklin, December 13, 1757. *B.F. Papers,* Volume 7, pp. 295–298; quoted in Van Doren, *Benjamin Franklin,* p. 272.

113 *His large secular works:* Handel arranged for a flock of sparrows to be released during the opera *Rinaldo.* His *Water Music* was performed from a barge floating on the Thames. *Music for the Royal Fireworks,* written in 1749 to celebrate the Peace of Aix-la-Chapelle, called for the firing of live cannon.

113 *"modern" music increasingly:* B.F. to Lord Kames, June 2, 1765; *B.F. Papers,* Volume 12, p. 162.

113 *screaming without cause:* B.F. to Peter Franklin, quoted in Sonneck, *Suum Cuiques,* 1916, pp. 81–82.

114 *The one musical composition:* See "Concert of Chamber Music by Benjamin Franklin and Wolfgang Amadeus Mozart," American Academy of Arts and Sciences, April 11, 1956; program in *B.F. Papers,* Yale University.

114 *Being charmed by the sweetness:* Quoted in Van Doren, *Benjamin Franklin,* p. 298.

114 *Concert of Musick:* Sonneck, pp. 59–60.

115 *The advantages of this instrument:* Quoted in Van Doren, *Benjamin Franklin,* p. 298. "The ear of a mortal can perceive in its plaintive tones the echoes of a divine harmony," commented Chateaubriand, while Goethe thought he heard in the armonica *die Herzblut der Welt*—the lifeblood of the universe.

115 *a Portable Instrument:* Advertisement in *London Chronicle,* June 17–19, 1762. Quoted in *B.F. Papers,* Volume 10, p. 118.

117 *a very easy afternoon:* Anecdote related in *B.F. Papers,* Volume 10, p. 119, footnote.

117 *an apt method:* Johann Friedrich Rochlitz in *Allgemeine Musikalische Zeitung,* 1798; quoted in Nicolella, "Glass Armonica: A Brief History."

117 *experienced nervous disorders:* B.F. to Giambatista Beccaria, July 13, 1762, *B.F. Papers,* Volume 10, pp. 116–130.

118 *This Mud being of a salt quality:* Clarence W. Alvord, editor, *The New Regime 1765–1767.* Springfield, Ill.: Illinois State Historical Society, 1916, p. 58; quoted in *B.F. Papers*, Volume 12, p. 399.

118 *He had no choice:* Croghan's description of his two trips to Big Bone Lick are in Simpson, "The Beginnings of Vertebrate Paleontology in North America."

118 *the Creature when Alive:* Letter from James Wright to John Bartram, August 22, 1762; in Simpson, "The Beginnings of Vertebrate Paleontology."

120 *too bulky to have the Activity:* B.F. to Jean Chappe d'Auteroche, January 31, 1768. *B.F. Papers*, Volume 15, p. 34.

120 *'Tis certainly the Wreck of a World:* See Campbell, *Recovering Benjamin Franklin*, p. 70.

121 *though we may as philosophers regret:* Hunter quoted in Simpson, "The Beginnings of Vertebrate Paleontology."

121 *fantastically precise misconception:* Gorst, *Measuring Eternity*, pp. 40–42.

121 *Franklin himself had published:* B.F. Papers, Volume 12, pp. 10–11.

122 *Join or Die:* The cartoon appeared in the *Pennsylvania Gazette*, May 9, 1754.

122 *It is better to be humbled:* Gorst, p. 106.

123 *This estimate:* Buffon, *Des Époques de la Nature*, p. 40.

123 *every thing leads us to believe:* Buffon quoted in Peale, *Account of the Skeleton of the Mammoth*, p. 13.

124 *Such is the economy of nature:* Jefferson, *Notes on Virginia*, p. 73.

124 *the Americans towered over the French:* Peale, pp. 43–44.

125 *there was not one American present:* Quoted in I. Bernard Cohen, *Science and the Founding Fathers*, p. 87.

126 *Since we had the misfortune: London Chronicle*, September 23–26, 1769.

126 *as a shower of stones:* Fonvielle, *Thunder and Lightning*, pp. 215–216.

128 *After a lightning rod was installed:* I. Bernard Cohen, *Benjamin Franklin's Science*, pp. 122–123.

128 *The proud Frederick:* Fonvielle, p. 277.

128 *When a chapel on Tottenham Court Road:* "An Account of the Death of a Person Destroyed by Lightning in the Chapel in Tottenham-Court Road, and Its Effects on the Building, as Observed by Mr. William Henly, Mr. Edward Nairne, and Mr. William Jones." *Philosophical Transactions*, Volume 62, 1772, pp. 131–136.

129 *On March 15, 1773:* "Account of the Effects of a Thunder-Storm, on the 15th of March 1773, upon the House of Lord Tylney at Naples. In a Letter from the Honourable Sir William Hamilton, Knight of the Bath, His Majesty's Envoy Extraordinary at the Court of Naples, and FRS, to Mathew Maty, MD, FRS." *Philosophical Transactions*, Volume 63, 1773–74, pp. 324–332.

129 *An examination the next day:* Arago, *Meteorological Essays*, p. 141.

129 *When the Purfleet committee:* "A Report of the Committee Appointed by the Royal Society, to Consider of a Method for Securing the Powder Magazines at Purfleet," *Philosophical Transactions*, Volume 63, 1773–74, pp. 42–48.

129 *By points we solicit the lightning:* "Report of the Committee . . . Powder

Magazines at Purfleet," *Philosophical Transactions*, Volume 63, 1773–74, p. 42.

130 *thunder is the greatest of electricities:* Wilson in *Philosophical Transactions*, Volume 54, 1764, pp. 249–251; quoted in Heilbron, *Electricity in the 17th and 18th Centuries*, p. 381.

130 *power of invitation:* Benjamin Wilson, "Observations Upon Lightning and the Method of Securing Buildings From Its Effects, In a Letter to Sir Charles Frederick, Surveyor General of His Majesty's Ordnance," *Philosophical Transactions*, Volume 63, 1773.

130 *In 1753 Beccaria:* Arago, pp. 230–231.

131 *Mr. Wilson's Objections:* "Purfleet Committee to the Royal Society, December 1772," in *B.F. Papers*, Volume 19, p. 425.

131 *daring "The Philosopher":* Benjamin Wilson, *Further Observations upon Lightning*, London, 1774, p. 15; quoted in *B.F. Papers*, Volume 20, p. 166.

131 *Pointed Conductors to secure Buildings:* B.F. to Saussure, Oct. 8, 1772; *B.F. Papers*, Volume 19, p. 325.

132 *Upon the whole:* B.F. to Lord Kames, Feb. 26, 1767; *B.F. Papers*, Volume 14, pp. 69–70.

132 *Would you kindly:* This anecdote has been reprinted in many forms. Franklin himself seems to have taken it from a 1714 volume by Robert Hunter titled *Androboros* that set the story, and the man with the hot iron, on one end of the Pont Neuf in Paris. See *Bulletin of the New York Public Library*, Volume 68, 1964, p. 153.

133 *prime conductor:* Cited in Isaacson, p. 277.

133 *It seems that I am too much:* Wright, "Benjamin Franklin: The Old England Man," p. 53.

133 *a bit muddled:* Private communication from Prof. R. W. Home.

134 *Mischievous Wilson:* Jean-Hyacinthe de Magellan's letter, written on September 15, 1777, was sent to another friend of Franklin's, Achile-Guillaume Leb que de Presle, with the request that it be forwarded to Franklin. Leb que de Presle did so on October 1, 1777. Benjamin Franklin Collection, Library of Congress.

134 *They are so plain:* Quoted in Heilbron, p. 382.

135 *I have no private interest:* Quoted in Schonland, *The Flight of Thunderbolts*, p. 30.

135 *This request the Society:* The Pringle incident is discussed in Heilbron, p. 382.

135 *While you, George:* Singer, "Sir John Pringle and His Circle," pp. 168–180.

136 *Physicist Philip Krider:* Private communication with author.

CHAPTER FIVE

137 *Silas Deane:* Deane was a former member of the Continental Congress from Connecticut, and Arthur Lee was a physician from Massachusetts who had studied law in London.

137 *I am like the remnant:* Rush, *Autobiography of Benjamin Rush*, p. 149.

138 *in "address and good breeding":* Quoted in I. Bernard Cohen, *Science and the Founding Fathers*, p. 175. John Adams also wrote of Franklin in Paris, "Newton had astonished perhaps forty or fifty men in Europe . . . Franklin's

fame was universal. His name was familiar to government and people, to kings, courtiers, nobility, clergy, and philosophers, as well as plebeians, to such a degree that there was scarcely a peasant or a citizen, a *valet de chambre*, coachman or footman, a lady's chambermaid or a scullion in a kitchen, who was not familiar with it, and who did not consider him as a friend to human kind." John Adams, letter to *Boston Patriot*, May 15, 1811.

138 *Bolstering his faith:* I. Bernard Cohen, *Science and the Founding Fathers*, pp. 175–176.

138 *Every Smith can make these:* "Memorandum on the Use of Pikes, Summer 1775," *B.F. Papers*, Volume 22, pp. 181–182.

138 *a Flight of Arrows:* B.F. to Charles Lee, Feb. 11, 1776; *B.F. Papers*, Volume 22, pp. 342–343.

139 *the project of enlightenment:* Gay, *The Enlightenment*, p. 558.

139 *Franklin, meanwhile, proved to be a delightful enigma:* Franklin's mission to France benefited from the actions of the Marquis de Lafayette, the young French nobleman who came to America in 1777, in defiance of Versailles, wintered with Washington at Valley Forge, and returned to France a popular hero in 1779. Franklin encouraged French public enthusiasm for Lafayette's virtuous pro-American example.

140 *happy mediocrity:* Benjamin Franklin, "Information for Those Who Would Remove to America" (1784) in *Writings*, edited by Leo Lemay, pp. 975–983; cited in Riskin, *Science in the Age of Sensibility*, pp. 73–74.

140 *Now one of the first characters:* European Magazine and London Review, Volume III, 1783, p. 163; quoted in Kraus, *The Atlantic Civilization*, p. 238.

140 *Seest thou a Man:* Proverbs 22:29; cited in Benjamin Franklin, *The Autobiography of Benjamin Franklin*, p. 144.

141 *visited Versailles "in the dress of an American farmer":* Schama, *Citizens*, p. 44. "The French king and queen remained perhaps the only people in that country not impressed by Franklin," Schama notes. "Louis XVI liked to putter in his workshop like a tradesman and Marie Antoinette enjoyed dressing up as a shepherdess, but both failed to see the charm and genius in the American who really was a common man." The queen, hearing of Franklin's preference for Gluck over Piccinni, a composer fancied by the court, is said to have asked, "What can a man whose trade is to put rods on buildings know about music?"

142 *Franklin's own popularity:* Schama, p. 43.

142 *more respect and veneration:* Thomas Jefferson to William Smith, February 19, 1791, Etting Collection, *Signers of the Declaration of Independence*, Volume 50, Historical Society of Pennsylvania; quoted in Hindle, *Pursuit of Science*, p. 223.

142 *He became used to saying:* Parton, *Life and Times of Benjamin Franklin*, pp. 434–435.

142 *The Franklin legend:* Maurois, *The Miracle of France*, p. 261.

143 *The two Aged Actors:* Adams quoted in *B.F. Papers*, Volume 26, p. 362, footnote.

143 *establish himself:* Voltaire quoted in *B.F. Papers*, Volume 25, p. 673, footnote.

143 *It is universally believed:* Quoted in Schama, p. 44.

143 *Dr. Franklin's electrical rod:* John Adams to Benjamin Rush, 1790. Quoted in I. Bernard Cohen, *Science and the Founding Fathers*, pp. 211–212.

144 *I can only suggest:* Lopez, *My Life with Franklin*, p. 175.

144 *I rise almost every morning:* B.F. to Barbeu Dubourg, July 28, 1768, *B.F. Papers*, Volume 15, pp. 180–181; also in Goodman, *The Ingenious Dr. Franklin*, p. 25.

144 *people often catch cold:* B.F. to Benjamin Rush, July 14, 1773, *B.F. Papers*, Volume 20, pp. 314–316; also in Goodman, p. 155.

144 *died "a Sacrifice":* Quoted in Ellis, *Passionate Sage*, p. 66.

144 *Somebody, it seems:* B.F. to Elizabeth Partridge, Oct. 11, 1779, *B.F. Papers*, Volume 30, p. 514.

145 *The temple of human nature:* John Adams letter to *Boston Patriot*, May 15, 1811.

146 *No, Monsieur:* Delaplaine's Repository, December 1818, Volume 2, pp. 114–115; see also Ramsey, *The History of the American Revolution.*

146 *[Mr. Adams] is always an honest Man:* Quoted in Lopez, *My Life with Franklin*, p. 175.

146 *It is true I have omitted:* B.F. to Arthur Lee, April 3, 1778; *B.F. Papers*, Volume 26, p. 223.

148 *By this means:* B.F. to George Whatley, May 23, 1785; in Benjamin Franklin, *The Writings of Benjamin Franklin*, Volume 9, pp. 331–339.

148 *if all the other Defects:* B.F. to George Whatley, August 21, 1784, in ibid., pp. 264–266.

148 *On August 27:* Quoted in Kerr, "What Good Is a New-Born Baby?," p. 27.

149 *There was a vast Concourse of Gentry:* B.F. to Joseph Banks, November 21, 1783; in *Writings*, Volume 9, pp. 113–118.

150 *Never before:* B.F. to Joseph Banks, November 21, 1783; in ibid.

150 *the first man to see:* Hamblyn, *The Invention of Clouds*, p. 112. On January 19, 1784, the first "passenger" flight took place, when seven intrepid souls ascended with Joseph-Michel Montgolfier in a balloon named "La Flesselles." On June 4, at Lyon, a "Madame Thible" became history's first woman to go aloft.

150 *A few Months [ago]:* B.F. to Joseph Banks, November 21, 1783; in *Writings*, Volume 9, pp. 113–118.

151 *Your Philosophy seems to be too bashful:* B.F. to Joseph Banks, November 21, 1783; in ibid.

152 *We should not suffer Pride:* B.F. to Joseph Banks, November 21, 1783; in ibid.

152 *where is the prince:* B.F. to Jan Ingenhousz, January 16, 1784; in ibid., pp. 155–156.

152 *a discovery big:* Jan Ingenhousz to B.F., January 2, 1784; Collection of the American Philosophical Society, Philadelphia.

152 *The delighted recipient:* The letter was actually from William Franklin to his son, William Temple Franklin. It is now in the American Philosophical Society Collection in Philadelphia. See *American Heritage*, December 1973, p. 28. Blanchard later that year staged an ascent in Philadelphia before a gathering that included George Washington. The initial balloon

craze came to a tragic end on June 15, 1785, when Pilâtre de Rozier, who had been the first man to fly, also became the first to die in a balloon. His craft caught fire as he attempted to repeat Jeffries' and Blanchard's Channel crossing in reverse.

154 *This was not the first time:* Quoted in Heilbron, *Electricity in the 17th and 18th Centuries*, pp. 364–365.

154 *a church steeple was struck:* The idea that divine wrath controlled thunderbolts became harder to dispute in the face of incidents like that at one European village in April 1760, when an abbey was reduced to rubble, having been struck by lightning three times within twenty minutes.

155 *dictated by ignorance:* Riskin, p. 149. My discussion of the Saint-Omer case is indebted to the description provided in Jessica Riskin's *Science in the Age of Sensibility: The Sentimental Empiricists of the French Enlightenment.*

155 *hostility toward the* monde savant: Riskin, p. 150.

156 *This is how to deal with:* Quoted in Vellay, *Robespierre et le Procès du Paratonnerre*, p. 28.

156 *There are some great lords:* Quoted in Henry, *Loss of Life in the United States by Lightning*, p. 17.

157 *shift for themselves:* Quoted in I. Bernard Cohen, *Science and the Founding Fathers*, pp. 169–170.

157 *Poor Buissart:* Riskin, p. 168.

158 *there was reason for the most enlightened person:* Quoted in Hamblyn, p. 66.

159 *Lightning accepted its laws:* Huet, "Thunder and Revolution: Franklin, Robespierre, Sade," pp. 18–19.

160 *Is man so unfamiliar:* Huet, pp. 19–20.

160 *Marat had observed:* See Vellay, pp. 206–207.

160 *Sir, A writ of condemnation:* Robespierre to B.F., October 1, 1783, Collection of the University of Pennsylvania Library; also see Vellay, p. 215.

162 *impressed by the marvelous:* Schaffer, "The Consuming Flame: Electrical Showmen and Tory Mystics in the World of Goods."

163 *imposing of mien:* Gillespie, *Science and Policy in France at the End of the Old Regime*, p. 261.

163 *[The] aggressively rationalistic:* A by-product of the reasonable eighteenth century, Castle explains, was the notion of the uncanny. As Freud would later suggest, reason can only suppress most of the time the primitive fears inculcated in the human experience. When an event or a sudden fear occurs that defies reason, man's latent memory of his most primitive beliefs comes rushing back, at times with overwhelming power. One of the best-known representations of this idea is the 1797 illustration by Francisco de Goya, *El Sueño de la Razón Produce Monstruos* (*The Sleep of Reason Produces Monsters*). A more playful variation were late eighteenth century magic lantern exhibitions, in which patrons in a darkened room marveled at magnified projections of insects and other recognizable scientific specimens, then became apprehensive or even fled in terror when the images became too apparition-like. Castle, *The Female Thermometer*, p. 8.

164 *Do you know that Herr von Mesmer:* Quoted in "Concert of Chamber Music

by Benj. Franklin and Wolfgang Amadeus Mozart," program notes, American Academy of Arts and Sciences, 1956. Other well-known composers who wrote for the armonica included Joseph Haydn, Ludwig van Beethoven, and Richard Strauss.

167 *I never knew any advantage:* B.F. to Sir John Pringle, December 21, 1757, *B.F. Papers,* Volume 7, pp. 298–300; Riskin, p. 196; see B.F. letter to Ingenhousz, April 29, 1785; in *Writings,* Volume 9, pp. 307–321.

168 *curative powers of electric fish:* In the early 1770s, attention turned from the torpedo to the large electric eels captured by ships' captains off the coast of Surinam and brought back live to Europe and the colonies for exhibition. Eels were found to have even larger shocking powers than the torpedo; one investigator claimed to have shocked thirty people at once with a jolt from an individual eel. The *Philosophical Transactions* published numerous papers on the subject, and in Philadelphia the American Philosophical Society pursued a series of experiments on eels in 1773 headed by Ebenezer Kinnersley and David Rittenhouse. See letters from John Walsh to B.F. in *B.F. Papers,* Volume 19, pp. 189–190, pp. 204–206, pp. 285–289; and Volume 20, pp. 258–267.

168 *temporary shadows:* Another French physician known for his medical electricity experiments was Jean-Paul Marat. Long before he attained fame and martyrdom as a journalist and pamphleteer in the French Revolution, the *"ami du peuple"* advocated medical electricity. In one notable experiment he cut open the chest of a live rat and sent electrical shocks through the creature's body to see how its heart and lungs responded. As historian Peter Heering suggests, Marat's belief in the efficacy of sudden electrical shocks to heal the body may have foreshadowed his political prescription for rousing the public. See Heering, "Jean Paul Marat: Medical Electricity Between Natural Philosophy and Revolutionary Politics"; in Bertucci and Pancaldi, editors, *Electric Bodies,* p. 91.

169 *quacks are the greatest:* "Conversations with Dr. Franklin." Benjamin Rush Manuscript Collection, Pennsylvania Historical Society, Volume 3, pp. 175–185.

169 *Still, electrical medicine:* Electrical medicine, known in the "spark and snap" period of the late eighteenth century for controversy, quackery, and miraculous expectations, resurfaced in the modern electrical age as an accepted form of medical diagnosis and treatment. X-rays, electrocardiograms, electroencephalograms, electroshock, and the pacemaker were among the many uses to which medicine put electricity, although, as in earlier times, extreme claims were sometimes made for its capabilities.

169 *It was the resulting gossip:* Lopez, *Mon Cher Papa,* p. 175. The one dissenting voice on the commission was the doctor and botanist Antoine-Laurent Jussieu, who thought the imagination alone could not account for the effects brought about in patients by mesmeric treatments.

170 *Never has a more:* Bailly, "Exposé des Expériences," report on the experiments which were conducted for the investigation of animal magnetism, read to the Academy of Sciences by M. Bailly, September 4, 1784. Printed by order of the king, at Paris, September 1784; quoted in Riskin, p. 205.

171 *There are in every great rich City:* B.F. to M. de la Condamine, March 19, 1784, in *Writings*, Volume 9, pp. 181–183.

171 *From the depths:* Joseph Michel Antoine Servan, a mesmerist and member of the Parlement of Bordeaux, quoted in Riskin, p. 193.

173 *If the young man had felt:* Bailly, "Exposé," pp. 43–45.

174 *If I had no country of my own:* Pierre Jean Baptiste Nougaret, *Anecdotes du Règne de Louis XVI*, 1791, Volume 4, p. 438; quoted in Aldridge, *Franklin and His French Contemporaries*, p. 202.

175 *On being first announced:* Martin Van Marum, "On the Theory of Franklin, According to Which Electrical Phenomena Are Explained by a Single Fluid." *Annals of Philosophy*, 1820, Volume 16, pp. 440–453; quoted in I. Bernard Cohen, *Franklin and Newton*, pp. 569–570.

177 *Even when his affairs kept him:* In Philadelphia in the 1750s, Franklin was involved in the sponsorship of two arctic voyages of exploration aimed at finding a Northwest Passage from Hudson Bay to the Pacific, a dream that would obsess mariners and gentlemen scholars for decades. Parliament in 1745 had offered 20,000 pounds to the first English subject to locate the Passage, and Franklin, himself a great hoaxer, was among those taken in by an account published by a dubious "Admiral De Fonte," who claimed to have made a trip along such a Passage traveling west to east. The two voyages of exploration from Philadelphia were on the ship *Argo*, captained by Charles Swaine, and were followed eagerly by Franklin and the *Argo*'s other backers. The first, in 1753, got only as far as Labrador before finding its way blocked by ice; a second attempt the following year turned tragic when several of its crew were murdered by Eskimos. The *Argo* returned to Philadelphia with only some native souvenirs to show for its effort, which were donated to the Library Company.

177 *As early as 1746:* See Alexander Small to B.F., April 13, 1772, *B.F. Papers*, Volume 19, pp. 105–110.

177 *We have informed them:* Folger's and B.F.'s letters are quoted in Gaskell, *The Gulf Stream*, pp. 5–7.

178 *river in the ocean:* This phrase describing the Gulf Stream was probably first used in the mid-nineteenth century by the pioneering American oceanographer Matthew Fontaine Maury.

179 *sundry very large Comazants:* "A Letter from Captain John Waddell to Mr. Naphthali Franks Merchant, Concerning the Effects of Lightning in Destroying the Polarity of a Mariners Compass," *Philosophical Transactions*, Volume 46, 1749–50, p. 111.

180 *the Electrical Fire . . . drawing off:* B.F. to Peter Collinson, June 29, 1751, *B.F. Papers*, Volume 4, p. 143.

181 *There was, unfortunately, great bureaucratic reluctance:* One reason for the Navy's resistance to Harris was probably its regret and public embarrassment, in 1824, over the installation of so-called "Davy protectors," a form of zinc sheathing meant to halt the corrosion of ships' copper-plated bottoms, designed by the British chemist Sir Humphry Davy, president of the Royal Society. While Davy's sheathing did, through an electrochemical process, inhibit corrosion of the copper, it also tended to pick up a far

greater number of weeds and barnacles that "fouled the ship's bottom," slowing its movement. Because of Davy's eminence, perhaps, or simply because at the time there was no adequate formal process of testing an innovation before implementing it, the Royal Navy had installed the zinc protectors on a large number of ships before anyone suspected there was a problem. Davy, who fancied himself a Franklinesque provider of useful science—he had earlier invented a widely used miner's safety lamp—saw his reputation blemished by the fiasco. See *London Times*, October 16, 1824; see also James, "Davy in the Dockyard: Humphry Davy, the Royal Society and the Electro-Chemical Protection of the Copper Sheeting of His Majesty's Ships in the Mid 1820s."

181 *For years, lightning continued:* In 1830, the Navy allowed thirty ships to be fitted out with pointed rods in their masts. Over the next fifteen years, these ships were completely unharmed by lightning, while forty-one others, not protected with Harris's device, were severely damaged. In summer 1839 the Royal Navy appointed a commission that ruled favorably on Harris's ideas, and in 1843 Harris at long last was authorized by the Navy to supervise the installation of lightning rods on ships' masts in all Her Majesty's dockyards. Knighted in 1847, he was ordered to oversee lightning protection for the new houses of Parliament and several other official buildings and palaces, including the royal mausoleum at Frogmore. Thanks to Harris—*and Franklin*—England's monarchs could repose eternally without fear of lightning.

182 *I promised to finish:* Franklin, "A Letter from Dr. Benjamin Franklin to Mr. Alphonsus le Roy . . . containing sundry Maritime Observations."

182 *employed merely in transporting superfluities:* Franklin, "A Letter from Dr. Benjamin Franklin to Mr. Alphonsus le Roy."

182 *Rejoice with me:* B.F. to Jan Ingenhousz, April 29, 1785; in *Writings*, Volume 9, pp. 307–321.

CHAPTER SIX

184 *Pure and exalted reason:* David Hartley to B.F., August 1789; Collection of the American Philosophical Society, Philadelphia.

184 *Electricity . . . may lead to:* Peter Collinson to Colden, March 30, 1745; Colden, *Letters and Papers of Cadwallader Colden*, Volume 3, pp. 109–111.

184 *O that moral science:* B.F. to Joseph Priestley, Feb. 8, 1780; *B.F. Papers*, Volume 31, pp. 455–456.

185 *knowledge and virtue were inseparable:* Riskin, *Science in the Age of Sensibility*, p. 71, and footnote; the Newton quote is from *Opticks*, Query 28.

185 *if natural Philosophy in all its Parts:* Quoted in I. Bernard, Cohen, *Science and the Founding Fathers*, p. 58.

185 *Everything has changed:* Robespierre, *Discours et Rapports à la Convention*, Paris, 1965, pp. 246–247; also in Huet, "Thunder and Revolution: Franklin, Robespierre, Sade," pp. 23–24.

186 *What Voltaire and his co-conspirators:* Robert Darnton, *George Washington's Teeth*. New York: W. W. Norton, 2003, p. 8.

186 *The sole foundation for belief:* Condorcet, *Oeuvres*, 1847, p. 404, in Huet, p. 21. Like John Adams, the historian Carl Becker thought Condorcet and the *philosophes* fatally misguided. "They renounced the authority of church and Bible, but exhibited a naïve faith in the authority of nature and reason. They denied that miracles ever happened, but believed in the perfectibility of the human race." Becker, *The Heavenly City of the 18th Century Philosophers*, pp. 30–31.

186 *A great man:* Condorcet, quoted in Gillespie, *Science and Polity in France at the End of the Old Regime*, p. 3.

187 *A new and intoxicating liberal idea:* Ellis, *After the Revolution*, p. 26.

188 *science of freedom:* The subtitle of the second volume of Peter Gay's *The Enlightenment.*

188 *Franklin meant to imply:* I. Bernard Cohen, "The Empirical Temper of Benjamin Franklin"; in Wright, *Benjamin Franklin: A Profile*, pp. 70–71.

188 *Science and its philosophical:* Rossiter, *Seedtime of the Republic*, p. 206, p. 213.

188 *philosophers had spoken:* Ibid., pp. 375–376.

189 *The sacred rights of mankind:* Hamilton, quoted in ibid., p. 376.

189 *angel in the whirlwind:* The title and epigram of Benson Bobrick's history, *Angel in the Whirlwind: The Triumph of the American Revolution.* New York: Penguin Books, 1997. The original quote is from a letter from John Page to Thomas Jefferson, dated July 20, 1776: "We know the race is not to the swift nor the Battle to the Strong. Do you not think an Angel rides in the Whirlwind and directs this Storm?"

189 *at an Epocha:* Washington, quoted in Rossiter, p. 377.

189 *The Arts have always traveled:* B.F. to Charles Willson Peale, July 4, 1771, *B.F. Papers*, Volume 18, pp. 162–163.

190 *The actual 1890 census:* Conway Zirkle, "Benjamin Franklin, Thomas Malthus and the United States Census," in Hindle, *The History of Science, Selections from ISIS.* Malthus's 1798 *Essay on the Principle of Population* shared Franklin's ideas about population but focused on some potential problems related to exponential growth, such as inadequate food supplies and the inability of "natural checks" such as famines, epidemics, and wars to maintain the population at a level the earth could sustain.

190 *Instinctively more comfortable:* Isaacson, *Benjamin Franklin*, p. 3.

191 *It is more than probable:* Quoted in Carr, *The Oldest Delegate*, p. 73.

192 *Thirteen staves and never a hoop:* Pelatiah Webster, quoted in ibid., p. 71.

192 *make short extemporaneous speeches:* Madison, quoted in ibid., p. 18.

192 *the wisest must agree:* Quoted in ibid., p. 76.

192 *not in associating:* Quoted in ibid., p. 67.

192 *By the Collision:* Benjamin Franklin, "The Internal State of America"; in Benjamin Franklin, *The Writings of Benjamin Franklin*, Volume 10, pp. 120–121.

192 *men with a nice feeling:* Rossiter in "The Political Theory of Benjamin Franklin" in Wright, *Benjamin Franklin*, p. 164.

193 *The convention did not need:* Carr, p. 58.

193 *There was, however, to be no compromise:* In somewhat marked contrast to "Franklinet," John Adams's son John Quincy had greater success distin-

guishing himself in small but significant diplomatic roles while acompanying his father abroad. In 1781, at age fourteen, young Quincy was already private secretary to Francis Dana, the American minister to Russia; in 1783 he served as the United States secretary at the Treaty of Paris that ended the American Revolution.

193 *She was going to a Brook:* Benjamin Franklin, "Queries and Remarks Respecting Alterations in the Constitution of Pennsylvania," November 2, 1789; in *Writings*, Volume 10, pp. 57–58.

193 *I have lived, Sir, a long time:* Benjamin Franklin, "Motion for Prayers in the Convention," June 28, 1787; in ibid., Volume 9, pp. 600–601.

194 *When the country's oldest Deist:* Berkin, *A Brilliant Solution*, p. 107.

194 *If God governs:* B.F. to Charles Dumas, January 18, 1781; in *Writings*, Volume 8, pp. 195–197.

195 *a nation of silk growers:* B.F. to Cadwalader Evans, September 7, 1769; *B.F. Papers*, Volume 16, p. 201. In the 1750s, the Society of Arts promoted silk cultivation in the Carolinas and Georgia by paying bounties on silk cocoons, although Pennsylvania made the most successful attempt, through the beneficence of the American Philosophical Society, which arranged for a silk filature and set prices. When Franklin was in London in the early 1770s, he oversaw some of the first sales of American silk. See Morgan, *The Gentle Puritan*, pp. 147–151.

195 *inventor of an early steamboat:* Franklin, for all his sagacity, seems to have underestimated the potential of steam power, a technology that would characterize industrial growth on both sides of the Atlantic in the coming decades. Indeed, the westward expansion of the United States he so earnestly anticipated would be largely driven by it. Fitch, whose early contributions to the development of the steamboat were largely ignored, and who always felt wronged by Franklin's indifference, later committed suicide.

196 *There was no curiosity:* Cutler, *Life, Journals, and Correspondence of Rev. Manasseh Cutler,* Volume I, pp. 267–269. Manasseh Cutler was a founding member of the American Academy of Arts and Sciences, a Boston-based scientific society (originally envisioned by John Winthrop) that John Adams helped create in 1780.

196 *I was highly delighted:* Cutler, Volume I, pp. 262–269.

196 *At one point Franklin began telling:* Cutler, quoted in I. Bernard Cohen, *Science and the Founding Fathers*, pp. 154–155.

197 *I confess that I do not entirely approve:* Farand, editor, *The Records of the Federal Convention of 1787*, Volume II, pp. 641–643. While the delegates were signing the document, Franklin remarked, "I have often, and often in the course of the session . . . looked at that behind the President [the sun painted on the back of the president's chair] without being able to tell whether it was rising or setting. But now at length I have the happiness to know that it is a rising and not a setting sun." This anecdote appears in Van Doren, *Benjamin Franklin*, p. 755, and elsewhere.

197 *Doctor, what have we got:* Quoted in Carr, p. 122. The Convention's end prompted another of Franklin's famous bons mots. In a letter to his old

friend, Jean Baptiste Le Roy, he said, "Our new constitution is now established, and has an appearance that promises permanency; but in this
world nothing can be said to be certain, except death and taxes!" B.F. to
Jean Baptiste Le Roy, Nov. 13, 1789; in *Writings*, Volume 10, pp. 68–69.

198 *Life, like a dramatic Piece:* B.F. to George Whitefield, July 2, 1756, *B.F.
Papers*, Volume 6, pp. 468–469.

198 *The Migration or Importation:* U.S. Constitution, Article 1, Section 9, paragraph 1.

198 *Franklin had been able to countenance:* Brands, *The First American*, p. 703.

198 *On Reflection it now seems probable:* Remark by B.F., Feb. 9, 1789, *B.F.
Papers*, Volume 5, p. 417.

199 *But such Mistakes:* Benjamin Franklin, *The Autobiography of Benjamin
Franklin*, pp. 211–212.

200 *the first abolitionist group in the New World:* The group had been founded in
1774–75.

200 *those persons, who profess to maintain:* Constitution of the Pennsylvania
Society for Promoting the Abolition of Slavery, published in *American
Museum*, April 23, 1789, pp. 388–389.

200 *one of the most important events:* Raynal, *A Philosophical and Political History
of the Settlements and Trade of the Europeans in the East and West Indies*,
Volume 3, p. 131, p. 140. This epic work made Raynal as famous as
Voltaire or Rousseau, and was said to have directly inspired the revolutionary fervor of the Haitian patriot Toussaint-Louverture. It was frequently reissued, going through no fewer than thirty editions between
1772 and 1789, despite an effort to suppress it and Raynal being forced
into exile in 1775.

200 *let, not only the deep sufferings:* Anthony Benezet to B.F., April 27, 1772, *B.F.
Papers*, Volume 19, p. 114.

201 *harsh and tyrannical treatment:* Autobiography, p. 69.

201 *he more than once made public:* "Remarks on Judge Foster's Argument in
Favor of the Right of Impressing Seamen," *B.F. Papers*, Volume 35,
pp. 491–502.

201 *In "Plain Truth":* "Plain Truth," *B.F. Papers*, Volume 3, pp. 180–204.

201 *by Nature a Thief:* "Observations Concerning the Increase of Mankind,"
B.F. Papers, Volume 4, pp. 225–234.

201 *The Whites who have Slaves:* Benjamin Franklin, "Observations Concerning
the Increase of Mankind," *B.F. Papers*, Volume 4, pp. 225–234.

202 *Deborah Franklin went to visit:* Franklin later recommended Newport,
Rhode Island; New York City; and Williamsburg, Virginia, to the Bray
Society as communities in which additional schools might thrive.

202 *rub on pretty comfortably:* B.F. to Deborah Franklin, June 27, 1760, *B.F.
Papers*, Volume 9, pp. 174–175.

202 *I was on the whole much pleas'd:* B.F. to John Waring, December 17, 1763,
B.F. Papers, Volume 10, p. 396.

202 *To Condorcet, a decade later:* B.F. to Condorcet, March 20, 1774, *B.F. Papers*,
Volume 21, pp. 151–152.

203 *Can sweetening our tea:* Benjamin Franklin, "The Sommersett Case and the

Slave Trade," published in the *London Chronicle*, June 18–20, 1772. *B.F. Papers*, Volume 19, pp. 187–188.

203 *I understood her Master:* Because of the Sommersett case, which had resulted from Granville Sharp's legal challenging over the status of slaves, an English court had ruled in 1772 that slaves became free upon arriving on British soil. B.F. to Jonathan Williams, July 7, 1773, *B.F. Papers*, Volume 20, pp. 291–292.

203 *a cancer that we must get rid of:* Charles Thomson to Thomas Jefferson, November 2, 1785, quoted in I. Bernard Cohen, *Science and the Founding Fathers*, p. 297.

204 *Logistical issues lent:* U.S. Bureau of Census, *First Census of the United States*. Baltimore: 1978, pp. 6–8.

204 *They have observed with great Satisfaction:* Benjamin Franklin, Pennsylvania Abolition Society Memorial to Congress, Feb. 3, 1790. Historical Society of Pennsylvania.

205 *evil spirits hovering:* DePauw, editor, *Documentary History of the First Federal Congress, 1789–1791*, p. xvii.

205 *It is well known the rice cannot be:* James Jackson, February 12, 1790, in ibid., p. 296.

206 *dirty pamphlets:* James Jackson, February 11, 1790, in ibid., p. 287.

206 *The pirate explained:* B.F. to the *Federal Gazette*, March 23, 1790; in *Writings*, Volume 10, pp. 86–91.

207 *Do these men expect a general emancipation:* Tucker, on February 12, 1790, in DePauw, p. 306.

207 *The people of the southern states:* Ibid., p. 308.

207 *Filibustering for two successive days:* Ohline, "Slavery, Economics, and Congressonal Politics, 1790," pp. 335–360.

207 *a matter of moonshine:* Fisher Ames in Ohline, p. 353.

208 *call us back to our first principles:* Ellis, *Founding Brothers*, p. 112.

208 *no doubt but his present afflictions:* Remarks of John Jones, M.D., in Pepper, *The Medical Side of Benjamin Franklin*, p. 130

208 *You have merited:* Ezra Stiles to B.F., January 28, 1790; Library of Congress.

209 *For my own personal Ease:* B.F. to George Washington, September 16, 1789; in *Writings*, Volume 10, p. 41.

209 *If to be venerated:* George Washington to B.F., September 23, 1789; *The Papers of George Washington*, University of Virginia Press, 1993, Volume 4, p. 66.

209 *circumstances have carried us too far:* Lavoisier to B.F., February 2, 1790, in a private collection, transcript at the University of Pennsylvania.

210 *the solid political education:* Maurois, *The Miracle of France*, p. 286, p. 289.

210 *Condemned to death:* Baker, *Condorcet*. p. 352.

211 *One of the final glories:* Gillespie, p. 144.

211 *I trust we shall never forget:* John Baynes, September 23, 1783; in *The Diary of John Baynes* (August 27–October 17, 1783), in *B.F. Papers*, Yale University.

211 *at length the invention:* B.F. to Professor Landriani, Oct. 14, 1787; in *Writings*, Volume 9, pp. 617–618.

212 *Indeed, in the construction of an instrument so new:* B.F. to Kinnersley, Feb. 1762, in *B.F. Papers,* Yale University, Volume 10, pp. 37–59.

EPILOGUE

213 *a revolution would occur:* Some 1,148 barns, stables, and granaries were reported struck and set afire by lightning in America in the three-year period 1890–92, compared with only 376 dwellings and 69 churches, according to an 1894 survey by the U.S. Weather Bureau. McAdie, *Protection from Lightning,* p. 8.

214 *a streak of real Jersey lightning: Scientific American,* September 5, 1885; a reproduction of Jennings's first lightning photo, taken on September 2, 1882, appears in *Journal of the Franklin Institute,* Volume 253, no. 5, May 1952.

215 *When the first photographs of lightning: New York Times,* February 27, 1927.

216 *In a typical thundercloud:* Golde, *Lightning Protection,* pp. 10–11. My description of these phenomena is indebted to Dr. Graydon Aulich, atmospheric scientist at the Langmuir Laboratory for Atmospheric Research, New Mexico Institute of Mining and Technology.

217 *If each storm is assumed:* Schonland, *The Flight of Thunderbolts,* p. 59, p. 99, p. 101; I. Bernard Cohen, *Benjamin Franklin's Science,* p. 155.

217 *may relieve us a hundred times:* B.F. to Ebenezer Kinnersley, February 20, 1762, *B.F. Papers,* Volume 10, pp. 37–59.

219 *form any of those diamonds:* Quoted in Stifler, *The Religion of Benjamin Franklin,* pp. 115–116.

219 *Good apprentices . . . are most likely:* Benjamin Franklin, Last Will and Testament, in *Writings,* Volume 10, p. 493; see also the pamphlet "A Sketch of the Origin . . . of the Franklin Fund," by Trustees of the Boston Franklin Fund, 1866. Each fund, Franklin calculated, would be worth £131,000 by 1890, and £4,061,000 by 1990, estimates that have proven relatively close, although due to the long-ago disappearance of the apprentice system, both funds have been adapted and used for various other civic purposes. In 1990, two centuries after Franklin's death, the Boston fund was valued at $5 million; the Philadelphia fund, managed less successfully, was worth $2.3 million. See Brands, pp. 712–713.

219 *produced a Fly:* Letter by Samuel Petrie, August 9, 1796, quoted in Joel J. Gold, "Dinner at Doctor Franklin's," p. 391.

220 *We may perhaps learn:* B.F. to Joseph Priestley, February 8, 1780, *B.F. Papers,* Volume 31, pp. 455–456.

220 *era of technological enthusiasm:* The phrase "era of technological enthusiasm" was popularized by Thomas P. Hughes in *American Genesis: A Century of Invention and Technological Enthusiasm, 1870–1970,* New York: Viking, 1989.

222 *We are all invited:* Quoted in Ellis, *Passionate Sage,* p. 202.

222 *Mr. Mesmer will be contented:* Quoted in Sonneck, *Suum Cuiques,* p. 71.

223 *tho' it is a question I do not dogmatize upon:* "Here is my creed," Franklin wrote to Ezra Stiles, president of Yale, who queried him in the last year of

his life about his religious beliefs. "I believe in one God, Creator of the universe. That he governs it by his Providence. That he ought to be worshipped. That the most acceptable service we render to him is doing good to his other children. That the soul of man is immortal, and will be treated with justice in another life respecting its conduct in this. These I take to be the fundamental principles of all sound religion." B.F. to Ezra Stiles, March 9, 1790; in *Writings*, Volume 10, pp. 83–85.

223 *a future State . . . all that here appears wrong:* B.F. to James Hutton, July 7, 1782; ibid., Volume 8, p. 562.

223 *I cannot suspect the Annihilation of Souls:* B.F. to George Whatley, May 23, 1785; ibid., Volume 9, p. 334.

INDEX

PHILIP DRAY is the author of *At the Hands of Persons Unknown: The Lynching of Black America*, which won the Robert F. Kennedy Memorial Book Award, the Southern Book Critics Circle Award for Nonfiction, and was a Finalist for the Pulitzer Prize and the *Los Angeles Times* Book Award. He lives in Brooklyn, New York.

ABOUT THE TYPE

This book was set in Caslon 540. The original typeface was first designed in 1722 by William Caslon. Due to its widespread use by most English printers in the early eighteenth century, Caslon soon supplanted the Dutch typefaces that had formerly prevailed. The roman font is considered a "workhorse" typeface due to its pleasant, open appearance, while the italic is exceedingly decorative.

Caslon's types also became popular throughout the American colonies. In his work as a printer, Benjamin Franklin made it his typeface of choice. The first printings of the American Declaration of Independence and the Constitution were set in Caslon.

The display face is Franklin Gothic, a typeface named posthumously in honor of the renowned printer, statesman, and inventor.